TECHNICAL

ILLUSTRATION

TECHNICAL

ILLUSTRATION

MATERIALS, METHODS AND TECHNIQUES

JUDY MARTIN

MACDONALD ORBIS

323211

AUTHOR'S ACKNOWLEDGEMENTS

Special thanks to Nigel Osborne for technical and editorial advice. Thanks are also due to Mark Way, John Fisher, Colin Rattray, Kevin Jones and Keith Harmer for time and interest given to the project and for supplying much useful information; to staff and students of the technical illustration departments at Portsmouth College, Ravensbourne College and Middlesex Polytechnic; to Amanda Baker for her thoughtful hard work; and to all other individuals and agencies who have provided information about their work in this field and/or contributed illustrations for reproduction in the book.

A **MACDONALD ORBIS** BOOK

© Macdonald & Co (Publishers) Ltd 1989

First published in Great Britain in 1989
by Macdonald & Co (Publishers) Ltd
London & Sydney

A member of Maxwell Pergamon Publishing Corporation plc

British Library Cataloguing in Publication Data
Martin, Judy, 1950–
    Technical illustration.
    1. Technical drawings
    I. Title
    604.2
    ISBN 0-356-17566-9

Filmset by Wyvern Typesetting Ltd, Bristol
Printed and bound in Italy by OFSA SpA, Milan

Editor: Gillian Prince
Designers: Clair Lidzey/Bob Burroughs
Picture Researchers: Amanda Baker/Helena Beaufoy
Photographer: Susanna Price
Illustrator: Michael Gilbert
Art Director: Linda Cole

Macdonald & Co (Publishers) Ltd
66–73 Shoe Lane
London EC4P 4AB

CONTENTS

▶ INTRODUCTION

It might at one time have been easy to begin a survey of technical illustration with a precise, and relatively narrow, definition of what constitutes a typical example of the genre. However, as technologically sophisticated products are increasingly absorbed into everyday life, there is a comparably expanding need for technical information presented in a form accessible to the lay person. One of the most efficient ways to convey such information is graphically — through pictures, diagrams and graphic symbols. Against this background, technical illustration has recently become a vital channel of communication, an area of graphic presentation which is being altered and extended in response to changing requirements.

WHAT IS TECHNICAL ILLUSTRATION?

Opinions might vary on this point: some illustrators will spend their whole careers working in a relatively restricted field both of subject matter and technique; others will encompass a wide range of subjects and may have the opportunity to try various different ways of presenting them visually. For the purposes of this book, the definition of technical illustration is taken quite simply to be illustration of man-made materials, objects and constructions, possibly including the situations in which they are used and the processes and systems in which they are incorporated. Technical subjects include vehicles and transport systems by air, land or sea, from space probes to the family car; working structures, from nuclear power stations to oil rigs to windmills; heavy industrial and agricultural machinery, from whole processing plants down to the smallest components of individual machines; telecommunications apparatus and networks; computer hardware; domestic appliances and everyday items such as cameras, calculators, radios and watches; and energy sources for mechanical and electrical devices, such as pump and motor actions, wiring and lubrication

systems. This is far from being an exhaustive list, but it indicates the general context of technical illustration work.

The main purpose of technical illustration is to describe or explain these items to a more or less non-technical audience. At one end of the scale, the illustrator is required to produce a visual image which is completely accurate technically and in terms of actual dimensions and proportions, to the point where someone can physically construct a technical object by reference to the illustration. At the other end, the purpose of the illustration is to provide an overall impression of what an object is or does, to enhance the viewer's interest and understanding.

Ideally, the illustrator has a dual fascination with how things work and

with the drawing and painting methods which can be used to render technical subjects. The work of different illustrators varies in style and presentation, but artistic licence is certainly limited as compared, say, to illustration work for advertising or fiction publishing. Technical illustration has an educational element which requires accuracy and attention to detail. The illustration has to carry some conviction; even where measured accuracy is not essential, the illustration must look right and contain all the information about the subject that the viewer needs in the given context.

Having described what technical

Above: Ford Sierra
Artist: Keith Harmer
Studio: Blue Chip Illustration
Client: Ford Motor Company Ltd
The cutaway is a high point of technical illustration, exemplified by this precisely rendered airbrushed image. This type of prestigious illustration work is used in new product promotions and sales literature. It identifies the overall shape and character of the car body but provides a view right through the engine and internal mechanisms. The continuing lines of the outer contour are shown over the open interior.

illustration is, it is also useful to define what it is not. There are related fields of scientific illustration in the areas of medicine and natural history, but these are considered to be distinctly separate disciplines. Architectural illustration, too, is a specialized subject area outside this field, though sometimes architectural aspects are incorporated in technical illustration. Science fiction illustration might benefit from the illustrator having technical knowledge, but fact is the domain of the technical illustrator.

The overall range of technical illustration will be seen by studying the examples of illustration selected for this book. This will make clear that a simple definition is not appropriate, in terms of content, style or technique. The selection is deliberately broadly based to encompass the many areas of work which the trained technical illustrator may encounter when entering professional practice. These areas are currently opening out, and there is some overlap with other graphic disciplines, as will be explained in succeeding sections.

THE CONTEXT OF TECHNICAL ILLUSTRATION

The original quite specialized context of technical illustration related to the practical application of technical information in industrial and manufacturing processes. For construction or servicing of mechanical objects, manuals were produced describing machine parts, assembly sequences and specific functions. The manuals included illustrations which presented the information explicitly in visual form, and these illustrations commonly derived from reference provided in plans drawn up by engineers and product designers, not from direct observation of a three-dimensional object. The special skill of the technical illustrator involved being able to read the blueprint information and render it into an image describing three-dimensional form, so that this image could then be related directly to the physical assembly of the object.

Descriptive visualization of newly developed objects and systems has

been an important aspect of technological development since the beginning of the industrial revolution. An illustration can always usefully supplement a verbal explanation, and sometimes replace words completely, bridging the gap between inventors and technicians on the one hand and, on the other, their interested audience previously equipped with little or no knowledge of the technical principles involved. Clear visual instruction on individual components and their assembly sequence was of crucial practical importance during World

Above: Pool table
Artist: John Scorey
Reproduced by kind permission of DPT Snooker Services Ltd
It is important to illustrate the solid construction of the pool table, but by exploding top and side the artist can also explain its interior functions. The continuation of surfaces from which parts have been exploded or cut away is traced with fine white lines. Airbrushing, finished with hand-painted detail, contributes the sleek surface quality of colour and texture.

of construction, servicing and maintenance, as well as promotional
presentations for new or particularly
prestigious products. This not only
provided the firm's staff, clients and
workers with necessary visual
material relating to different aspects
of the output, but also set up an
industrial training ground for technical illustrators.

CURRENT PRACTICE

In the past two decades, the situation
has again changed. Many illustrators
now work individually on a freelance
basis, gaining commissions either
directly from industrial clients, or sub-

Below: Autosampler for liquid
chromatography
Artist: Goro Shimaka
Client: System Instruments
Company Ltd
**A highly detailed and relatively
small-scale piece of technical
equipment is a particular challenge
to the illustrator's craft. Initially, an
accurate drawing must be achieved
as the basis for colour work. Careful
attention to subtle variations in the
neutral tones and highlighting effects
defines the different qualities of, for
example, glass and metal.**

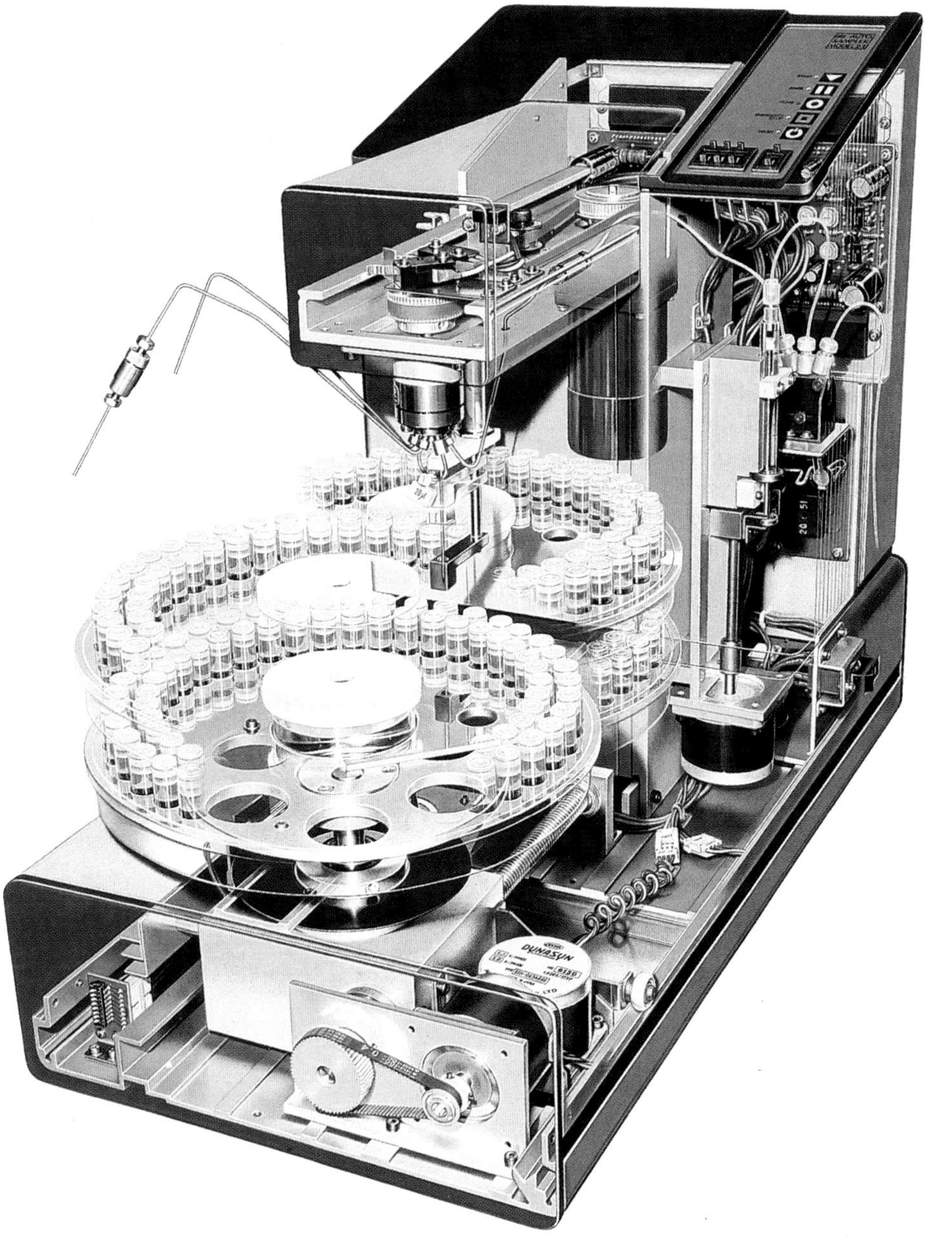

War II, for example, when technical
illustrations enabled unskilled workers to play an efficient part in
manufacturing, replacing the trained
workforce which was depleted by
conscription to the armed services.

In the post-war period, technical
illustration became more highly
developed through work in the automobile and aerospace industries.
During the 1950s and 1960s, it was not
uncommon for large industrial companies to house a permanent drawing
studio which dealt with the requirements for illustrative work in the areas

contracted through design groups handling the graphic presentations required by major companies. There are still some salaried positions in industry and in, for example, government defence departments, but industrial studios geared to training and utilizing specialist artists are no longer the norm. Some small independent studios have been set up by freelance illustrators, gathering three or four artists together to offer the particular skills of technical illustration.

Another element which has had a significant impact on the development of technical illustration during the 1980s is the higher profile of design services and a greater appreciation among manufacturers of the need for well-styled presentation of a product. This has produced an increasing amount of work in the areas of sales promotions and advertising. As products sold into the general marketplace become more technologically sophis-

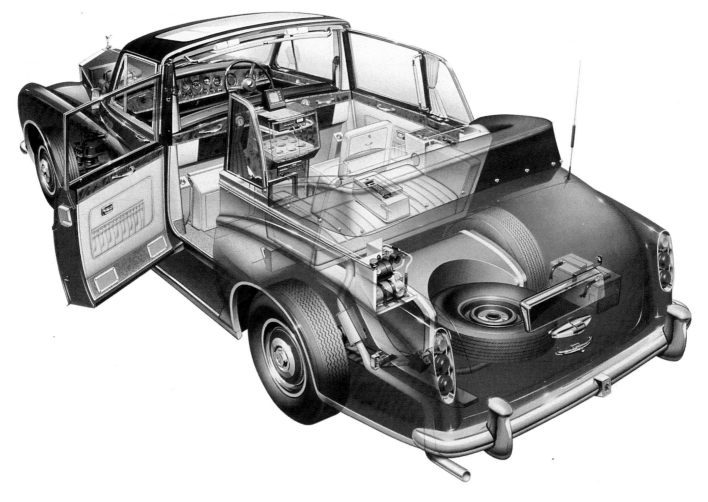

Above left: VW engine block
Artist: Timothy Connell
Portsmouth College of Art, Design
and F.E.
The need for descriptive detail in a technical image does not preclude a vigorous approach to the illustrative technique. This gouache and watercolour rendering shows a bold painterly style and an inventive association of line and solid colour.

Above right: Dumper truck
Artist: Peter Visscher, Aircraft
Client: Michelin Tyres plc
Transparent colour enhances detailed rendering of tonal values. Line work and colour washes are used here to create a crisp, graphic image with textural interest.

Left: Rolls-Royce Phantom VI State Landaulette
Studio: Warwickshire Illustrations Ltd
© Rolls-Royce Motor Cars Ltd
The technique of ghosting—showing interior parts through an outer casing—is highly developed in relation to automobile illustration. The organization of line detail and colour describes both the bodywork and interior components at various levels through the car.

ticated, there is an increasing need to provide some information about them to consumers, who both want and appreciate the products but have no technical knowledge about how they are constructed or how they work. Market research has shown, for example, that car buyers are attracted by the inclusion in sales brochures of highly finished airbrushed illustrations showing the interior structure and working parts of the vehicles. This increases consumer confidence, although the information conveyed in such an illustration cannot necessarily be 'read' technically by the potential buyer.

In relation to sales and marketing, technical illustration serves a purpose which cannot be achieved by other illustrative means. Despite widespread use of photography in promotional work and advertising, it is sometimes necessary to prepare visualizations of a new product before it is actually constructed, and this cannot be done photographically. A technical illustrator who can read engineering drawings and discuss details of the new product with the industrial engineers has a valuable role to play

in presenting a realistic image of the as yet unrealized item to its potential users. This approach is more direct and effective than relying on verbal descriptions, or plan documents which the future buyer cannot fully understand.

A broader field of work has opened up to technical illustrators during the past two decades with the growth of illustrated non-fiction publishing — educational and 'how-to' books, practical manuals, partworks and specialist magazines. The readership of these books and magazines has a considerable appetite for finding out how things work and how they, the readers, can achieve practical skills. Some of the publications deal with complex scientific and technical information explained for the lay person; others are designed to be of directly practical value, in areas such as home improvements and car maintenance, for example. Technical information is illustrated in a variety of ways, and has brought to a wider, non-specialist audience some of the graphic devices which have long been the stock-in-trade of the technical illustrator, such as exploded views and cutaways

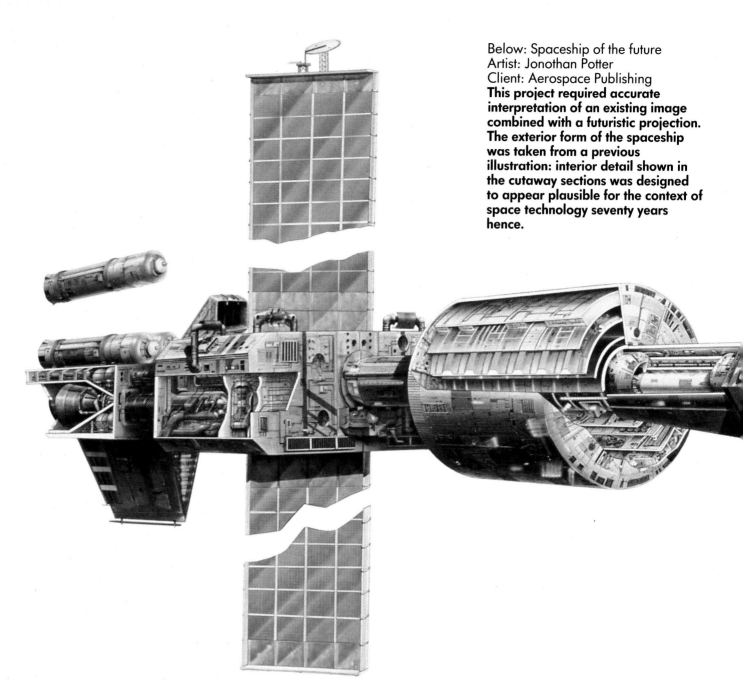

Below: Spaceship of the future
Artist: Jonothan Potter
Client: Aerospace Publishing
This project required accurate interpretation of an existing image combined with a futuristic projection. The exterior form of the spaceship was taken from a previous illustration: interior detail shown in the cutaway sections was designed to appear plausible for the context of space technology seventy years hence.

which reveal the internal workings of mechanical and composite objects.

There are two other notably specialized areas in which technical illustration is occasionally required. One is display work for museums, for permanent displays or special exhibitions. In major institutions, this work is usually planned and prepared by an in-house design team, but some illustration work is contracted to freelance artists. Such work may offer the chance to render technical subjects on a much larger scale than is

normally required for illustrative material. For large display panels, the illustrator may be expected to complete the work on site. Exhibition display work is an interesting, if rare, opportunity. An even more restricted field is illustration to be incorporated in special effects for filming. This is not the kind of commission which will be offered to freelance illustrators working more broadly in the technical field: the film industry trains its own artists in the particular techniques and visual effects required.

TRAINING AND PROFESSIONAL PRACTICE

With the changing circumstances of professional practice in technical illustration, studio apprenticeships are no longer the main site for training illustrators in the drawing and painting skills required for technical work. Specialized courses are run by some art schools and colleges of higher education. These equip the illustrator with the necessary expertise in the practical aspects of producing an illustration, and supply the technical background. The harsh reality of acquiring commissions, negotiating fees, working to a specific brief and meeting tight deadlines is only encountered in commercial practice: business skills are equally necessary to a successful freelance career but these have to be learned through practical experience of the market-place.

Typically, the content of a course in technical illustration puts considerable emphasis on developing drawing skills. Because of the complexity of many technical subjects and the requirement for a realistic representation, a detailed drawing is the essential basis of an illustration. Subsequent work on developing the style and presentation of the image is almost secondary, as highly developed skills in, for instance, airbrush painting are wasted if applied to an image which is faulty from the start. Objective drawing is emphasized, encouraging accurate observation which is the key to accurate rendering. Freehand drawing skills and drawing construction methods are extensively taught, including perspective systems and translation of orthographic (two-dimensional) information into three-dimensional form. At the same time the student is introduced to the technical context of the work, learning to identify specific components, understanding basic engineering methods, and investigating the actual construction of objects by taking things apart to have a look.

Drawings developed into full illustrations exercise skills of rendering in line — inking a drawing — and full colour, including hand-painting and airbrushing techniques. A different aspect of the work, but one which is of equally practical though sometimes less direct significance to the illustrator, is acquiring knowledge of the broad context of graphic design within which the illustrative content is placed. This includes a working

Right: Technical pen cutaways © Staedtler (UK) Ltd
An interesting contrast in two cutaway sections of the technical pen: the fully rendered version shows the visible construction of the pen nib unit and barrel: a diagrammatic rendering includes the passage of the flow of ink from reservoir to pen tip.

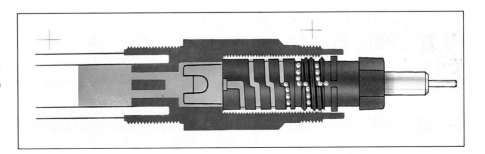

knowledge of typography and photography, as well as development of a general design sense, and the practical requirements of preparing an illustration for reproduction. Like any specialized area of work, graphic design has its own working methods, equipment and jargon, but this is all simply learned by experience.

A relatively new skill which the technical illustrator is now required to take on board is a knowledge of computer graphics functions, so that these can be put to use as and when they are available and appropriate. It might seem that technical work is the ideal area for computers to take over graphic presentation, but the days of the computer as technical illustrator are as yet far off. Currently, a fully rendered image created by computer is either an inaccessible or a far too expensive option for design groups and companies needing technical

illustration for their presentations. Programming systems which specifically answer the problems of producing technical illustration by computer are also still at the development stage. What is of value in the immediate future, however, is the use of computers to plot complex drawings and provide basic constructions from different viewpoints and angles. This can relieve some of the tediously methodical planning of an image which has to date always been part of the technical illustrator's lot.

A period of training and private study is a great opportunity to investigate the possibilities of technical illustration in terms of drawing and painting techniques and interpretation of subject matter. The great difference between studying technical illustration and going into practice as an illustrator, which hits home when the student goes out into the world, is the pace

and pressure of the work under commercial conditions. Especially when freelancing, the illustrator, although part of a team, is probably working in isolation when actually executing the illustration. Self-discipline and a professional approach both to deadlines and to fulfilling the details of a briefing are essential.

While it is to be hoped that the client or designer responsible for the briefing will welcome the expert contribution which the illustrator can make to

'Development of the steam engine' project
Artist: David Manchip
Portsmouth College of Art, Design and F.E.
A valuable function of computer art is the ability to generate progressive images with technical and graphic consistency. This is from a set of four illustrations created on the 'Pluto' Graphic Art Workstation.

the overall project, it is often the case that all the boundaries of a particular piece of work are laid down in advance and the illustrator's task is simply to follow instructions. Deadlines often seem impossible: learning to achieve the impossible gives the individual illustrator a head-start over colleagues and rivals. Fees vary according to the context of the work — advertising typically pays more than publishing, for example —

but in any job a budget for work contracted to freelancers will have been set initially in the costings and the illustrator must find out the parameters for negotiation of payments with each separate client.

Working conditions also vary. When producing a parts and servicing manual for direct use in industry, the illustrator usually collaborates with a technical author. Between them they work out the necessary information

Sprocket wheel
Artist: Tom Steyer, The Garden Studio
The purely functional elements of an unremarkable item of this kind can be surprisingly beautiful when described with the close attention to subtle colour detail provided by Tom Steyer in this rendering. Unusually, it is painted in oils which contribute their particularly rich density of colour.

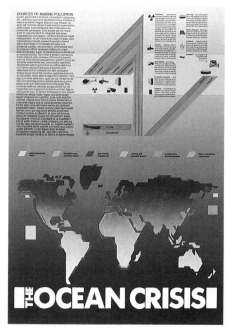

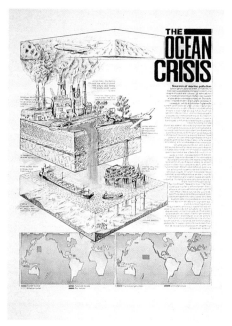

'The Ocean Crisis' project
Artists: (top left) Simon Adamson, (top right) Mark Ansell, (above left) John Vollands, (above right) Timothy Connell
Portsmouth College of Art, Design and F.E.
There are always a number of solutions to the design of a specific project, as shown in these four interpretations of an ecological theme. Two artists have opted for bold graphics, the others for a more organic 'hand-drawn' style.

and the best way to present it. In work for a small industrial company, the illustrator may be answerable directly to the client who is commissioning and paying for the work. In publishing, the editor or art director of the publication in production will usually provide the illustrator's brief. In sales, advertising and promotional work, the illustrator may be selected and briefed by a design group or advertising agency which is responsible for the style and execution of the graphic presentations.

The briefing sets out the subject and context of the illustration and states whether it is line or colour and how the visual style can be interpreted, as well as the practical details of deadlines and payment. It is usual practice for the illustrator to present a fully worked drawing before inking or colour work are applied, for approval by the client and checking by technical staff where necessary. Full reference for the information to be included in the illustration may be supplied by the client, but if it is the artist's responsibility to research the project, this should be reflected in the time-scale and fee for the work.

There is considerable competition for any type of illustration work, especially in the freelance market, and competence is not necessarily enough to promise a continuous flow of interesting work. Some artists gain an advantage by specializing in a particular subject area, some by developing a highly distinctive style; others do well out of versatility, being prepared to accept a wide range of work which increases their experience of commercial practice and opportunities in the field of illustration work generally. It should be noted that glamorous imagery such as airbrushed cutaway views of a new automobile represents a small percentage of available work, but more often than not, the technical illustrator is required to take a less creative role.

AN INTRODUCTION TO TECHNICAL ILLUSTRATION

This book cannot fully simulate a four-year technical illustration course or a studio apprenticeship. It aims to introduce the range of the subject in terms of the artistic skills which the illustrator must acquire, the special conventions of technical illustration and the broader fields of work which the trained illustrator may choose to enter. It is intended to provide a practical grounding in the techniques of technical illustration and, through the work of a number of successful practitioners in this area of illustration, a demonstration of what can be achieved and the ways in which it can be accomplished.

Montgomerie Reid fork lift truck
Artist: Colin Brown, The Garden
Studio
© Industrial Art Studio
**The delicate quality of airbrush spray
and use of transparent inks as the
medium allow a complex ghosted
image to be developed with great
clarity and precision.**

MATERIALS AND EQUIPMENT

The materials of the illustrator range from the most long-established of artists' media to new products specially developed for the current requirements of practising artists and designers. Since it is frequently the case that technical illustration requires a precise and controlled form of rendering, certain materials and tools are more commonly used than others for producing finished illustration work. However, the fields of work open to the technical illustrator are expanding and each new project has its own problems and solutions. This section examines the full range of materials and equipment appropriate to varying styles of illustration: their applications are comprehensively demonstrated throughout the illustrated examples of techniques and finished work in this book.

PENCILS

This most basic item of equipment is still one of the essentials of illustration. Pencils are used for sketching, developing an idea through rough drawings, and for drawing up the detailed construction of an image before applying black ink line or colour work.

Traditional wood-cased 'lead' pencils (the 'lead' is in fact a mixture of graphite and clay) are inexpensive and versatile tools. The degree of hardness of the pencil lead depends upon the proportion of graphite to clay in its composition, and is denoted by a B (soft) or H (hard) number — for example, 6B produces very soft, grainy, heavily black marks; 2H is a moderately hard pencil producing a fine grey-black line, and so on. Degrees of hardness vary from 9B to 10H, but not all grades are available in every manufacturer's range. (Some ranges include E and F codes: E pencils are very soft, while F lies between H and B qualities.) The higher numbers denote extreme characteristics and are probably inappropriate for most drawing and illustration work. The range from 2B to 2H covers most general needs, with harder grades for drawing up line work.

Mechanical pencils

Clutch and mechanical (also called propelling) pencils are more recent, and now much favoured, developments of the basic lead pencil. Each consists of a plastic or metal barrel encasing a sleeve which holds the replaceable lead. In the clutch pencil, a push-button mechanism releases the clutch tip, allowing the lead to be adjusted in length. In the mechanical pencil, which takes finer leads, the gripping mechanism is inside the barrel and the lead length is automatically adjusted by push-button control. The protruding section of the lead is protected by a very fine metal sleeve, leaving only the working tip exposed.

A standard size of lead for the clutch pencil is 2mm, though some types will take up to 3.15mm. These

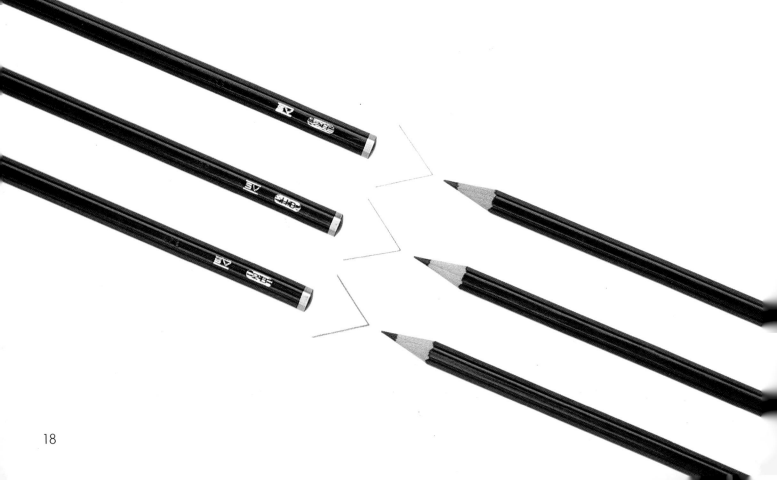

wood-cased pencil which shortens as it is used and sharpened, and the consistency of line and tone which the standard leads allow, useful for working up a detailed and accurately measured construction drawing. However, for freehand drawing and working sketches, the wood-cased pencils have a pleasing sensitivitiy and variety in their mark-making capabilities, which enables the artist to develop subtle qualities of form and texture in a drawing.

Blue lead (non-reproducing) pencils
Blue lead does not show up in photographic reproduction, which means that construction lines and working detail do not have to be erased from the final image after a drawing has been inked. Blue lead comes in wood-cased pencil form or as individual leads for insertion in clutch or mechanical pencils.

COLOURED PENCILS
The 'lead' in a coloured pencil basically consists of pigment bound in china clay. Wood-cased coloured pencils include three different types: those with heavy, soft coloured leads

require sharpening during use; a special sharpener is usually incorporated within the pencil casing, or a sandpaper block can be used to refine the point.

Mechanical pencils can be fitted with very fine leads from 0·3mm through 0·5mm, 0·7mm and 0·9mm. These do not need to be sharpened and are kept in constant use simply by

extending the working tip as necessary. Thicker leads may be used in some types, up to 3mm. There are also polymer-based hard leads made for use on drafting film, usually graded on a numerical range from one (soft) to five (hard).

The advantages of clutch and mechanical pencils are the uniform shape of the barrel, as against the

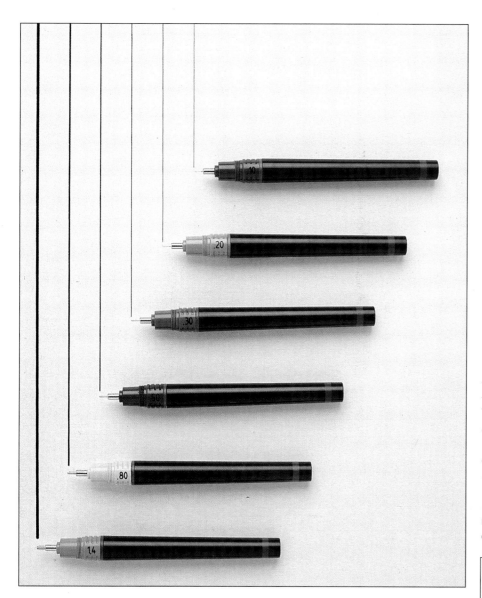

nib and the inconstant flow of ink to the pen tip, though they do produce expressive qualities in rough drawings and sketches. The most widely used type of pen for illustration and design work is now the technical pen, which has a fine, tubular nib and an in-built reservoir of ink. The nibs are supplied in standard sizes ranging from 0.1mm up to 2mm. There are two standard grading systems for nib width: each offers a range of very fine and consistently accurate nibs incorporated in a nib-unit casing, usually with its own reservoir attached. The entire nib assembly must be changed to vary line width. The nib size is marked on the unit casing and sizes are also colour-coded.

For the serious illustrator, a set of technical pens with different nib sizes is essential equipment, allowing maximum versatility for minimum effort. As these are relatively expensive items, the initial outlay has to be considered. The pens are sold in boxed sets, but an alternative is to buy two or three providing contrasting line widths, adding others to the collection as necessary.

The advantage of the technical pen is the uniform line quality and thickness which contributes a high degree of accuracy to line work. It may take a

which quickly wear down in use; those with finer, non-crumbly leads, which are most generally suitable for illustration work in this medium; and leads composed of water-soluble colour which can be spread into controlled brush marks or broad washes using a wetted paintbrush.

For working drawings and colour renderings of technical subjects, it is important to choose a type of coloured pencil that can be sharpened to a fine, durable point for line work but that also creates an even spread of soft colour when used for shading. Water-soluble colour is particularly useful where subtle colour-blending is required, but this can also be done by overlaying hatched linear strokes of different colours, which creates a

characteristically textured colour effect. The number of colours required depends upon the technique and style of work of the individual illustrator. The various manufacturer's ranges of coloured pencils offer from 24 to 72 colours; pencils may be available singly as well as in boxed sets.

Coloured leads are also available for some types of clutch pencils, and are handled technically in the same way as the wood-cased leads, that is, used for line work, colour shading and hatching.

DRAWING PENS
Ordinary dip pens are not considered suitable for technical illustration work as they are somewhat unpredictable in use due to varying pressure on the

while to become accustomed to working freely with a technical pen, as it needs to be kept at a consistent and upright angle, especially when used in conjunction with a ruler or stencil. The reservoir is filled with specially formulated non-clogging ink that dries quickly and is waterproof and smear-proof when dry, thus allowing pencil marks on a line illustration to be freely erased. Inks are available in black and a range of colours, supplied in small bottles with a nozzle extension which facilitates refilling of the pen reservoir. The nib unit of a technical pen is a sensitive element and the pen should be kept clean and left capped when not in use.

The range of Graphos pens consists of a holder incorporating an ink-feed, which can be fitted with any of a number of interchangeable nib units each designed for a different aspect of design and illustration work — fine or broad ruling, tubular, round and slanted nibs for technical and freehand work. This type of pen to some extent combines the qualities of the old-fashioned dip pen and the technical pen, offering a useful degree of technical versatility.

RULING PENS

The ruling pen is a technical instrument for drawing straight lines with paint or ink. The working end consists of a curved blade and a flat blade connected by an adjustable screw. The screw adjustment fixes the gap between the blades and this space forms the reservoir for the medium which flows through the tip of the pen, producing a consistent line width. This is a useful and economical tool for applying line detail to colour work. It needs careful control, but the skill is easily mastered with a little practice. A double nib attachment is also available for ruling parallel lines.

BALLPOINT AND FIBRE TIP PENS

There is now a huge variety of pens with synthetic fibre tips made for general purpose writing and drawing. They may contain water-soluble or permanent ink and the colour range covers all shades of the spectrum. Fibre-tip pens are an easily portable

source of colour for sketching, and are useful for working up an idea in colour before taking it to a finished stage in a different medium. They can also be used to add linear detail and sharp-edged qualities to marker or watercolour work: in this case, the quality of the ink should be checked beforehand, as water-soluble inks may lift under a paint or marker overlay. Spirit-based permanent ink is more versatile, but the pens must be kept capped when not in use, as the ink evaporates quite quickly if the tip is left exposed.

Ballpoint pens include inexpensive types with plastic barrel and metal nib, typically available in black, blue, green and red, and more sophisticated designs in which the barrel is shaped like a traditional writing pen. Ink refills are available for these models. The drawback with ballpoints has always been the variable quality of the ink supply, which tends to blob or streak from time to time. However, many artists favour this informal medium for sketches and working drawings as it encourages a quick, bold drawing technique.

Included in this category are nylon-tipped rolling ball pens — also available in a limited colour range — in which the ink flow allows a more even and sensitive line and has less tendency to smudging.

MARKERS

The potential of markers for roughs and finished work has recently been exploited much more fully than in the past. For illustrations and presentation renderings, high-quality studio markers are required. The principle is the same in all: the barrel contains a central ink reservoir supplying a felt-

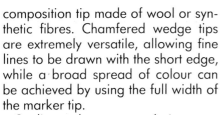

composition tip made of wool or synthetic fibres. Chamfered wedge tips are extremely versatile, allowing fine lines to be drawn with the short edge, while a broad spread of colour can be achieved by using the full width of the marker tip.

Studio markers are made in a vast colour range, usually referenced by naming and number-coding each colour; in some cases the colours are coordinated to papers, films and printing inks, allowing consistent colour matching in all areas of design and illustration work. Marker ink is water- or spirit-based: spirit-based is more versatile from the point of view of overlaying colour layers, but some types do bleed through and across the paper. The main drawback of markers is, however, the fact that the colours are not lightfast. Although they are excellent for short-term purposes such as presentation and reproduction, the illustration does not

have a long life in its original condition.

The ideal set of markers offers colour consistency throughout the range, so that hue and tone do not vary in a single application, or when a used marker is replaced with a new one. A marker should be easy to handle, making fluid lines and blocks of colour, and preferably bleed-resistant, so that the artist does not have to make allowances for colour spread on the support, when the marker ink creeps over outlines or into adjacent colour areas.

Any marker will quickly dry out if left lying around uncapped. As they are relatively costly items, it is as well to take care of them. Solvents are available for revitalizing dried markers, but these may affect colour consistency and should not be relied on when producing finished work, though revived markers can still be used on sketches and rough layouts. A

relatively new development has produced markers in which the ink is released from the reservoir through a valve only when pressure is applied to the tip, so that the ink supply is no longer vulnerable to exposure when the marker is not in use.

Fine-tipped markers are available with pointed nibs made from bonded

fibres; these can be used in conjunction with the larger wedge-tip types and are supplied by some manufacturers in a corresponding colour range. Again, these should be tried out for colour consistency.

It is useful to keep a colour chart coded by the names or numbers of the different colours in your marker range, applied to the surfaces which you will mainly use for marker rendering. Although manufacturers include colour flashes on the casings, because of the way these are printed it is impossible to make them absolutely accurate, and the surface tint and relative absorbency of a paper or board further affects the colour application.

DRAWING INKS
Apart from inks supplied for use in technical pens, there is a wide range of drawing inks, in black and a variety of colours, which can be used for various styles of illustration. Non-water-soluble inks consist of pigments or dyes in suspension in a liquid vehicle. They dry quickly to a durable, waterproof finish with a high degree of light-fastness. The colours are vivid and have a transparent quality that allows them to be overlaid to form subtle shading and colour-mixing effects. Black India or Chinese ink forms a solid, glossy black at full strength and can be diluted with water to provide grey washes for monochrome drawings.

Waterproof drawing inks typically contain shellac, a resinous ingredient that is hard to shift when dry. This type of ink should never be used in technical pens as it is likely to cause clogging. If it is used in an airbrush, the instrument must be thoroughly cleaned as soon as the work is completed.

An alternative for airbrush work is the range of water-soluble inks which are similar in performance to liquid watercolours. These are also of unsuitable consistency for use in technical pens, since they are insufficiently viscous and are likely to flood the nib.

Writing ink is an interesting medium for line-and-wash drawings, as it separates into irregular patches of colour when diluted with water. Standard black writing ink, for example, forms grey-blue washes tinged with coppery brown. The main drawback to this technique is that the ink is water-soluble and the clarity of the line can be lost when washes are applied.

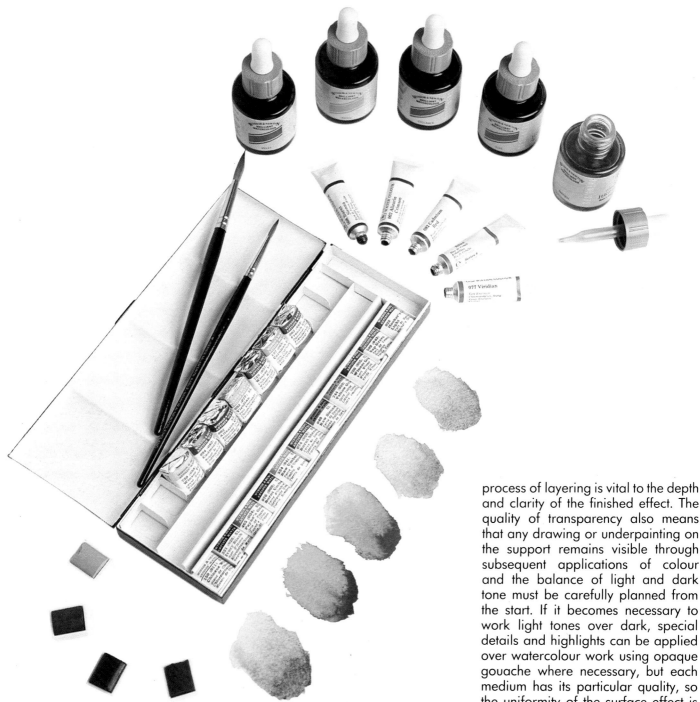

WATERCOLOUR

Watercolours consist of finely ground pigments bound in gum arabic. The traditional medium is available as dry cakes or semi-moist pans of colour, and in thick, fluid consistency packaged in tubes. Liquid watercolours are also available which are particularly well suited to airbrush work: the container usually includes a dropper for transferring the liquid directly from the bottle to the reservoir of the air-

brush, or into a palette for colour mixing.

Watercolour is a widely used and versatile medium for illustration. The colours have brilliance and transparency when they are applied to a pure white support, and can be built up layer over layer in thin washes, brushed or sprayed, which create subtly vibrant colour effects. Colours are extremely luminous when wet, but this quality fades as they dry, so the

process of layering is vital to the depth and clarity of the finished effect. The quality of transparency also means that any drawing or underpainting on the support remains visible through subsequent applications of colour and the balance of light and dark tone must be carefully planned from the start. If it becomes necessary to work light tones over dark, special details and highlights can be applied over watercolour work using opaque gouache where necessary, but each medium has its particular quality, so the uniformity of the surface effect is disturbed.

Various brands of watercolour are available in colour ranges of from 24 to 80 colours. Certain pigments tend to be fugitive, that is, they fade relatively quickly on exposure to light, and the product label may show a grading for the degree of permanence. A wide variation of tints and shades can be obtained by diluting colours with water to a greater or lesser degree. Distilled water is often recommended for this purpose as it

transparent than others, and the hiding power of the range of yellows, for example, is typically less than that of reds or blues.

Some illustrators work wholly in gouache while others reserve it for detail and highlighting on watercolour or marker work. It also combines well with pastel and with line media, pencil or ink.

Paints described as 'poster' or 'powder' colours are similar in formula to gouache, but they are not suitable for technical illustration. They are intended to be inexpensive products for general use, and the pigments and paint texture are not of sufficient quality for work required to a high standard of finish.

PASTEL AND CHARCOAL

Sticks of pastel are made by mixing pigment with a binder and compressing it into square- or round-sectioned lengths. The hardness of the pastel depends upon the proportion of binder in the mixture. There are three basic grades — hard, medium and soft. Some types are paper-wrapped, others given a fine plastic coating, to make them cleaner to handle. Pastel is

maintains the purity of colour.

In addition to washes, watercolours can be used to paint delicate detail, using fine sable brushes to hatch together strokes of pure colour, or for drybrush techniques of forming broken colour. The medium adapts easily to either a loosely worked or hard-edged style of image, and to mixed-media effects. Pencil or ink drawings can be given colour detail using watercolour washes, and interesting effects can be obtained by combining this fluid paint medium with the linear emphasis of coloured pencil or with the rich texture of pastel.

GOUACHE

By contrast with watercolour, gouache is a solidly opaque medium. The paint consists of pigment mixed with a white filler in a gum binder — the filler causes the opaque and slightly chalky surface finish. The colours are clear and brilliant but tend to lighten as they dry, so this should be taken into account when judging the tonal balance as an illustration progresses. Because of the opacity, it is possible to work light over dark or vice versa, though with some colours there is a slight risk of bleed-through because of the intensity of the colouring matter used in manufacturing the paint.

Various brands of gouache —

which may also be marketed as 'designer's colour' or 'tempera' — are available in colour ranges of up to 80 hues and shades. Like watercolours, these may be marked on the label with a colour permanence rating, and also with the degree of opacity. Some colours are by nature more

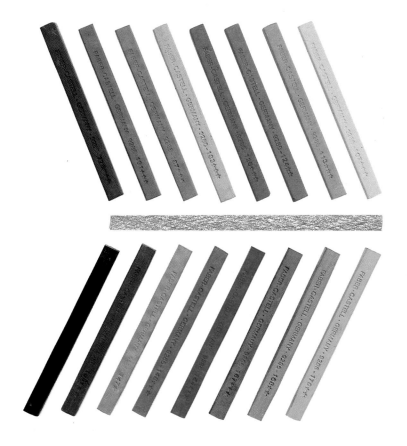

also manufactured as fine sticks encased in wood, like conventional pencils.

Colour ranges vary, but in the finest artist's pastels up to 144 hues, tints and shades are available, sold singly or in boxed sets. Light and dark tones of one colour are number-coded from 0 (palest) to 8 (darkest) shade. This graded range allows a subtlety of colour and tone to be created by hatching and overlaying linear strokes, but pastels can also be blended by rubbing with fingers, cotton wool, a brush, or with a torchon (stump), a tight paper roll with a shaped tip.

Pastel has a characteristically rich texture unlike any other medium, but it takes time to become skilled in controlling the spread of powdery colour. The stick form enables it to be used

like other drawing mediums, but it has been traditionally regarded as a paint medium because of the colour quality, with the advantage that there is no drying time to be considered. The support, paper or board, needs to have a distinct grain or 'tooth' to grip the loose colour, and the drawing must be sprayed with fixative at frequent intervals to avoid smudging.

Charcoal is associated with pastel since it is often used to create a keyline drawing over which the pastel colour is worked. It is produced by charring fine sticks of wood under controlled heat conditions and, like pastel, it is a loose, crumbly medium which needs fixing for permanence. Charcoal is a vigorous, bold medium for sketching and may be used to capture strong qualities of line and tone in

reference drawings, but it is rarely much employed in finished illustration.

Few illustrators work entirely in pastel, but the medium is also useful for adding final details of texture and highlighting over work mainly executed in marker, watercolour or gouache.

ACRYLICS

Acrylics and polymer-based paints are relatively recent additions to the range of artist's materials, but are now available in different forms and quantities to suit particular purposes. Standard acrylic paints sold in tubes or tubs are of a viscosity which imitates oil paint; there are also flow-formula types, which are more liquid in consistency, and designer's acrylics packaged similarly to tube watercolours and gouache.

Acrylics can be applied in layers of flat colour, as washes or glazes, with hatched or broken brushstrokes, or as thick impasto. They dry rapidly to a flexible and durable waterproof finish. They are available in a wide range of colours, having been developed in association with the expanding modern range of synthetic pigments and dyes. There is usually a degree of transparency to the paint quality, but acrylics can be worked layer upon layer indefinitely, as they do not pick up underlying colour from previous applications and dry quickly enough to be overworked within a very short period. They can be applied to almost any surface, though they do not adhere to an oil-based underlayer. Various media such as matt or gloss glazes and heavy gels are available for use with acrylics. They modify the working qualities of the paint and the effect of the surface finish.

Acrylic is an extremely versatile medium, but certain drawbacks accompany the advantages. The tough, waterproof skin formed by the paint cannot be lifted once it has begun to dry: corrections must be made by overpainting. Acrylics also tend to have a slight sheen when dry, due to the characteristic texture of the paint. The quick-drying property also generally makes the medium less suit-

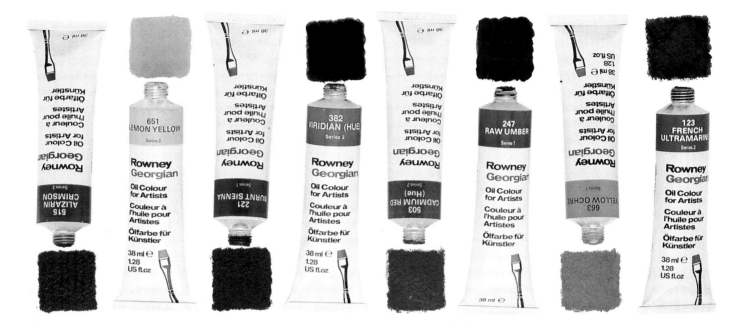

able for airbrush work than water-colour or gouache, as there is a greater risk that the paint will clog the airbrush nozzle and it is not easy to clean out once dried within the delicate instrument. Proprietary brand solvents are available for cleaning an airbrush or a hair or bristle brush that becomes clogged with dried acrylic, but nonetheless the dried paint can cause damage to the tool.

The waterproof quality of acrylic is of advantage in airbrush work when used to make a correction to a water-colour or gouache rendering. Spraying an opaque layer of white acrylic leaves a flat, clean surface that can be reworked completely and will not allow any colour to bleed through from underneath. A number of artists do prefer the qualities of acrylic, finding them well suited to their own styles of airbrush work. The essential rule is always to clean an airbrush immediately and thoroughly if acrylics are used.

OIL PAINT

The primary medium of painters for five centuries, oil paint is still more closely associated with fine art than with illustration (although this distinction is not always appropriate, since many original illustrations match fine-art paintings in quality). Oils have a vibrancy of colour and surface texture

that cannot be achieved in any other medium, and this lends itself to colour reproduction. However, the enormous disadvantage of oil paint for the working illustrator is its drying time. It remains workable for days and may still be drying out after a period of weeks. With tight deadlines to meet, it is not usually possible for an illustrator to make allowances for this property, whatever the fineness of the finished effect. Surfaces for oil painting also require lengthy preparation as compared to other media, unless the original illustration is to be regarded as completely disposable, and this hardly makes sense if a lot of work has gone into the painting.

This being said, there is a place for considering use of oils for prestigious finished images, where time and money allow the indulgence. In technical illustration, oil painting techniques are better suited to broadly realistic renderings, rather than diagrammatic or intricately constructed technical work.

The wide colour range of oil paints is complemented by the numerous mediums and varnishes which can be added or finally applied to enhance the paint quality. Oils can be worked as glazes, skins of colour, or thick, textured impasto. The rule is to work 'fat over lean', that is, increasing the heaviness and oiliness of the paint

layers successively. Heavy under-painting can destroy the finish if it dries more slowly than the top layers and distorts the surface quality.

Paper, wood panel, stretched canvas or prepared canvas board can be used as the support; all need to be properly primed and grounded before colour is applied. Suitable diluents and solvents must be used to soften and clean paintbrushes during and after use with oils.

BRUSHES

For work in watercolour and gouache, which are the main painting media used in technical illustration, brushes should be soft but resilient for applying and spreading colour evenly, with a fine, fluid point for detailing. Round sable brushes precisely fit these requirements and are the best choice for general studio work. They are available in sizes from 00 (very fine) to 12 (large, but still finely tipped). The number of brushes and size range you will require depend upon the type of work and usual working scale. As sable brushes are relatively expensive, it is advisable to start with three or four of suitable sizes and add to the selection as necessary. Sable brushes are also obtainable with flat, squared tips — useful for applying colour thickly and also for working into hard-edged angles — and shaped as soft fans for blending wet colour.

Ox-hair brushes also have good qualities of springiness and fluidity for laying colour, but do not provide such a sleek point for fine work. Although sable hair is favoured, synthetic hair brushes are now produced to very good quality and are more affordable for the beginner. These are also recommended for work in acrylic. For large-scale work in acrylics and oils, bristle brushes are commonly used, since they are stronger for working a viscous medium. Hog hair is the traditional material, but again the synthetic equivalents have been greatly improved in recent years.

Though brushes are made and shaped for particular functions there are no rules governing their use: the choice is extremely personal. If a brush feels comfortable and you are confident in handling it, that is the main recommendation. Avoid brush tips which are unevenly shaped or spreading, or are poorly fitted into the metal ferrule and liable to shed hairs into the paint. Always clean brushes thoroughly immediately after use, restore the proper shape and store them tips upwards.

THE AIRBRUSH

There is a variety of airbrushes on the market, with most manufacturers presenting a range of models, but all operate on the same basic principle: separate paint and air supplies are channelled through the body of the airbrush towards the nozzle where, under pressure, they are combined and expelled as a fine spray of colour. The number of different models is not, however, just a matter of presentation, as the various types have particular capabilities.

For detailed illustration work it is essential to have the type known as 'double-action', which means that paint and air supplies are activated separately by the user. The air supply is started by pressing one finger down on the control button; then the paint is released by pulling back on the button. With practice, the combined operation is smoothly achieved. The most versatile airbrushes have independent double-action, allowing the user to vary the proportions of paint and air; in fixed double-action airbrushes the ratio is preset. More skill is required to control the independent action consistently and achieve easy variation of the paint/air ratio when required, but this type of airbrush offers the illustrator by far the greatest range of airbrushing capabilities.

The paint reservoir may be a recess in the body of the airbrush, a metal cup mounted above or to one side of the nozzle, or a jar screwed into the top or bottom of the airbrush shaft. The medium is supplied to the airbrush in one of two ways: by gravity feed, in which the reservoir stands above the airbrush and the medium flows into it freely; or by suction feed, in which the medium is drawn up into the airbrush by the pressure loss

occurring as the air supply passes through above it. The position and size of the paint reservoir affect the handling of the airbrush: a large jar holds more medium than a recessed reservoir or integral colour cup, but its weight affects the balance of the instrument when it is held in the hand. There are one or two minor drawbacks and specific advantages to each of the various types, and the choice must be a matter of personal preference.

The air supply

The only reliable source of air for the airbrush is an electrically run compressor which channels a constant supply of air into the instrument via a connecting hose. An alternative is to use a can of pressurized liquid gas, connected to the air hose by a valve attachment. Being portable, these cans are useful if the illustrator needs to do a limited amount of work away from the studio, and they are an inexpensive way for the beginner to try out the capacities of the airbrush before investing in a compressor. But air cans do not provide either a consistent or a sufficient supply of air for sustained airbrush work.

In a design or illustration studio where airbrushing forms an important quotient of the day-to-day work, there will be a large compressor with several air outlets so that more than one airbrush can be used at the same time. This type of compressor incorporates a tank of air kept at constant pressure: as the air is fed into the airbrushes, the compressor reactivates to replenish the tank and maintain the pressure. It will also have a filter, a moisture trap to eliminate condensation that could disturb the spray quality, and a pressure regulator that allows different levels of pressure to be preset.

The individual illustrator can buy a small storage compressor that incorporates all these features, but as it is a less powerful machine, it may need to be rested at regular intervals during a prolonged period of work. There are other, even smaller compressors on the market which provide a more basic facility. The disadvantage of a

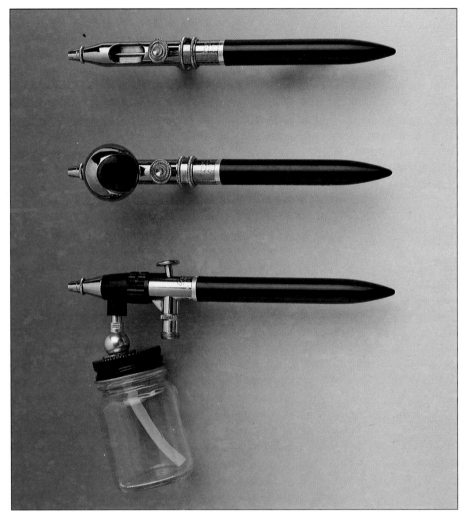

small compressor that does not have a storage tank is that the motor action translates into the airbrush spray as a slight pulsing motion, robbing the operator of a consistent spray quality.

Media for airbrushing

The most commonly used media for airbrush illustration are inks or liquid watercolours, both of which provide transparent layers of colour, and gouache, an opaque medium which allows the illustrator to overspray corrections and work light over dark colours. Liquid media can be transferred direct from the container to the paint reservoir of the airbrush. Gouache is a viscous substance that must be diluted with water to a milky consistency before use. Make sure no lumps of colour remain in the diluted mixture which could become trapped at the tiny nozzle opening at the tip of the airbrush. Thick gobbets of paint

either block the flow of medium or are spat on to the artwork, spoiling the smoothness of the sprayed colour. Acrylics can be used in the same way as gouache, but they dry quickly and are difficult to remove if allowed to dry inside the airbrush. When working with water-based media, clean the airbrush by flushing through with water until no further colour tint is visible. This procedure must be followed each time the colour is changed, and at the end of a working session.

Oil paint is not recommended for airbrushing, but if an oil-based medium is used, the tool must be well flushed out with the appropriate solvents to ensure all traces of paint are removed. Airbrush spray creates a heavy atmosphere within the studio, due to the gradual build-up of tiny particles of paint in the air. This is particularly unpleasant with oils, but it occurs whatever type of paint is used and can be dangerous. Make sure the room is well ventilated while you are airbrushing. Wearing a mask is advisable during continuous spraying.

Maintenance

The airbrush is a precision instrument and must be kept thoroughly cleaned, ensuring that no colour has accumulated either on the needle running through the centre of the airbrush which conducts the paint to the nozzle, or within the nozzle itself. It is not good practice to rely on scraping out colour that has been trapped and allowed to dry inside the airbrush, as rough handling affects the carefully constructed balance of parts which ensures the consistently fine spray quality. When disassembling the air-

brush for cleaning and maintenance, handle each part with care and place the components securely on a flat surface in correct order for reassembly.

Masking materials for airbrushing

Skill in manipulating the airbrush is only half the battle in airbrush illustration; the other half is a comparable skill in planning and cutting masks. Because of the 'cone' of spray created by the airbrush, it creates a soft effect and there is a slight area of overspray which blurs edges and dividing lines. To obtain hard-edged shapes, a masking material is needed to limit the spray to the precise area delineated by the artist.

A mask is, in simple terms, anything which blocks the path of the airbrush

spray, so that no colour gets through to the support. The most commonly used material is masking film, a transparent, self-adhesive plastic film which is tough but thin and flexible. It is supplied in sheets or rolls, with a backing paper to protect the adhesive side. The masking film is laid over the whole area of the image to be coloured and the mask sections are cut and lifted in sequence as the colour is applied (see pages 122–124 for masking techniques). Masking film has a special low-tack adhesive so that it can be lifted cleanly from paper or artboard without damage to the surface. Coloured films are available: it should be borne in mind that these interfere with the artist's perception of the colours being applied and should only be chosen if it is not particularly important to have a clear view of what is underneath the mask for matching of colours and tones.

A transparent low-tack tape can be used to remask small areas or to anchor loose masks. A loose mask can be made from any material — paper, cardboard, fabric, acetate, or even a three-dimensional object. The mask may form a positive or negative shape: for example, you can spray around a paper square or into a square hole cut into a large sheet of paper. The different shapes and textures of loose-masking materials provide a range of airbrushed effects.

Masking fluid

A liquid rubber solution that dries to a thick, non-porous skin, masking fluid is a useful material for masking out very fine details or creating drawn or brushed textures. It can be used with hand-painting techniques as well as in airbrushing, and may be applied with a fine brush or a ruling pen. The fluid is allowed to dry, the colour is applied over it, and when that is dry, the rubbery skin is peeled or rubbed away.

Masking fluid should be tested before it is used on work in progress, as some types leave a faint yellow stain on the surface of the support. This is a disadvantage if the purpose of the mask is to leave a white highlight area, but less crucial if the masked area is to be oversprayed.

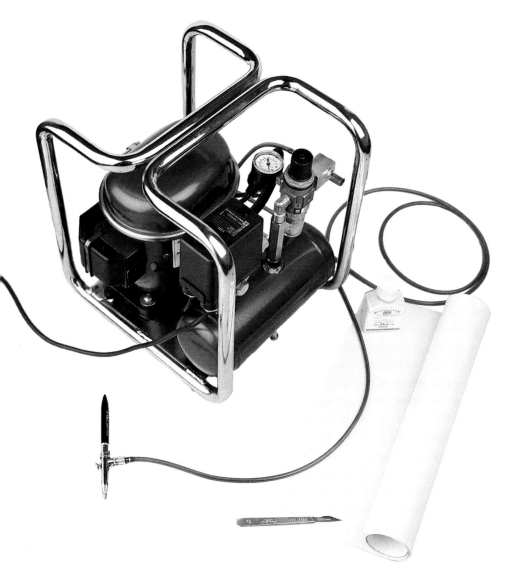

PAPERS AND BOARDS

The illustrator chooses a suitable support — that is, the surface to which an illustration is applied — taking into account the purpose of the work, the medium and the required qualities of surface finish. There is now an immense range of papers and artboards for illustration work covering all kinds of weights and textural qualities, and a broad price range.

Papers

For general drawing and sketching, including objective drawing and working out constructed forms, layout paper and cartridge paper are good for work in pencil, ink, marker, watercolour, gouache and pastel. They have the advantage of being inexpensive, so different construction methods and techniques can be tried out without the uncomfortable feeling of wasting a valuable material. Layout paper is pure white and translucent, allowing a drawing to be built up using underlays in successive stages. It is available in rolls or pads in standard A sizes. Marker pads also

▶ STRETCHING PAPER

1 Cut one strip of gummed paper tape for each side of the paper, allowing a little extra length than the actual measurement of the paper edge. Put the tape strips to one side and damp both sides of the paper using a wet sponge. Be careful not to wet the tape at this stage.

2 Spread the paper flat on a drawing board. Dampen one of the lengths of tape with the sponge and lay it along one edge of the paper. Smooth it in place so that it adheres firmly and evenly to the edge of the paper and the board.

3 Repeat the process to form an adhesive 'frame'. Allow the paper and tape to dry out thoroughly, keeping the board flat, before using the paper for painting. When the paper is wetted out with paint it will buckle, but the taped edges ensure that it dries flat again.

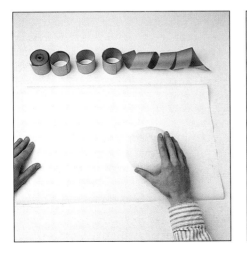

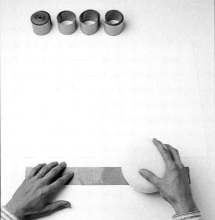

contain fine, lightweight white paper, but this is made bleed-proof for colour roughs. Cartridge paper comes in different weights and thicknesses, in white and colours, all having an even, relatively smooth surface. It can be obtained as pads, sheets or rolls.

Watercolour papers fall into three basic categories of surface finish: hot-pressed (HP) are smooth-textured papers; 'not' ('not hot-pressed', also called 'cold-pressed') papers have medium-rough surfaces; rough papers, as the name implies, have more grainy and heavily textured finishes. HP and 'not' papers are also suitable for work in gouache, but the flat finish of gouache is more at odds with the noticeable texture of the heavier rough papers. These papers are available in different weights, and all but the heaviest need to be stretched on a drawing board before paint is applied. Otherwise the paper sheet will buckle when wetted by the medium. Prepared watercolour boards are also marketed, as are blocks of paper sheets in various sizes. Watercolour papers are not suitable for line work using a technical pen, as the surface fibres obstruct the movement of the fine pen nib and tend to absorb and spread the ink, causing a ragged line.

Coloured papers range from lightweight sheets coated with bright colour on one side only, to the more subtle integral tints of Ingres paper, a grainy material favoured for pastel drawing. The stronger colours may have a use in providing a flat background tone for diagrammatic work, while tinted papers are sometimes preferred for work in pencil, pastel or gouache to set a mid-tone from which the lighter and darker tonal values are keyed.

Tracing paper and drafting film
Tracing paper, finely grained and transparent, is used not only for transferring the outlines of a drawing from one surface to another, but also for initial working out of full views and details. Working on tracing paper, each successive development in the drawing can be achieved by using the

previous version as an underlay, from which the important details are traced, gradually discarding construction lines and sketched-in details that are no longer vital to the finished image. Individual parts of an image drawn on tracing paper can also be assembled to produce a composite drawing, from which a final, clean version is derived.

Detail paper — also called 'typo detail' because it is used by designers for planning typographic layouts — has qualities in common with both tracing and layout papers, and can be used in the same ways. It is tougher than tracing paper, with a smooth, white finish, but slightly more transparent than layout paper, giving a clearer view of underlays.

Drafting film has recently become an important option for finished line work. It eliminates the need for a transfer method, as it can be laid over a completed drawing to allow the art-

ist to produce a cleanly inked line drawing following the lines of the drawing underneath. It accepts ink well, producing a clear, consistent line. The clear plastic typically has a matt finish on one side and is slightly glossy on the other. Some types have a blue tint which is non-reproducing. Drafting film is also used for overlays on which areas to be printed with colour tint are indicated separately from the keyline drawing.

All these materials are available in single sheets or pads conforming to standard international paper sizes.

Artboards
For line work and airbrushing, the surface needs to be extremely smooth and resilient, allowing an even application of the medium and the ability to scratch back ink or colour to make corrections without damaging the surface. Line boards and line-and-wash boards have these qualities and are

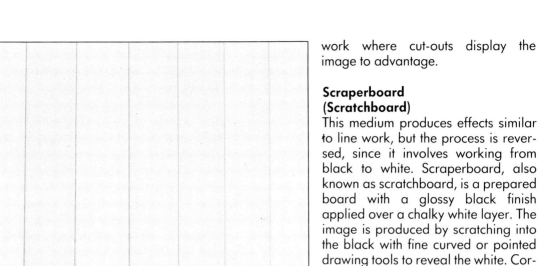

work where cut-outs display the image to advantage.

Scraperboard (Scratchboard)

This medium produces effects similar to line work, but the process is reversed, since it involves working from black to white. Scraperboard, also known as scratchboard, is a prepared board with a glossy black finish applied over a chalky white layer. The image is produced by scratching into the black with fine curved or pointed drawing tools to reveal the white. Corrections can be made by applying black India ink.

Because this medium is not commonly used, its effect is rather striking when it is successfully employed for illustration work. Technical illustrators skilled in line work using pen and ink may appreciate the different qualities achieved in working with this negative process.

DRAWING AIDS

In the nature of technical illustration, precision is an important aspect of the work, so drawing aids feature more largely in this field than in other areas of illustration. Drawing instruments such as compasses and dividers, measuring equipment and templates are all part of the technical illustrator's stock studio equipment. Careful thought should be given to the most essential items for a particular area of work, as covering every possible need would be an expensive prospect and although many useful and inventive products are available, few are indispensable tools.

Technical instruments

It is worthwhile investing in high-quality technical instruments, since they offer the only guarantee of precision. Making do with cheap versions is a false economy: they simply do not do the job. A good set of technical instruments includes a spring-bow compass — the two arms separated by a screw thread which fixes the required radius — with drawing leads to be inserted; a small-radius compass; an extension bar for drawing large circles and arcs; a ruling pen

specially designed for illustration work. The surface is brilliantly white, smooth and non-grainy.

In colour work intended for reproduction, a flexible support is normally required, as scanning methods involve fixing the artwork to a drum. In the past, the surface layers might be stripped from artboard to isolate the upper layer bearing the image, always a delicate process. A type of artboard can now be obtained in which the surface layer is bonded to the board by means of a self-adhesive backing, remaining stable while the artwork is prepared, but easily removed for reproduction processing. Alternatively, the finish of an artboard such as CS10 is available in paper weight. The surface quality is the same as that of the board, but the material is

lightweight and flexible.

Other useful artboards include illustration boards, which consist of high-quality watercolour paper bonded to a cardboard backing, offering a range of surface finishes, and Bristol board, a smooth white board available in various thicknesses according to the number of layers bonded together. Bristol board can also be used on either side, whereas most illustration boards have only one working side. These are all suitable for work in the main illustration media. For line work, drafting film is increasingly favoured (see page 32).

Foam-core boards consist of smooth white paper coating both sides of a rigid polystyrene board. This material can be cut accurately to shape and is useful for presentation

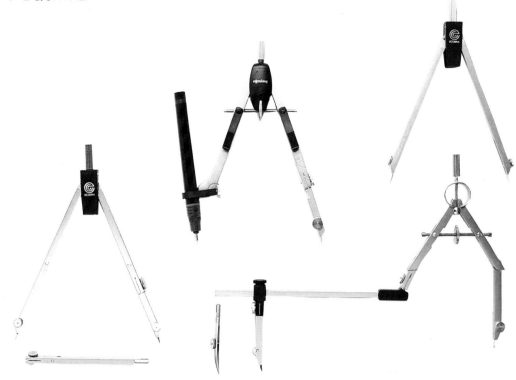

attachment for the compass; and dividers. The advantage of buying a boxed set is that the cushioned box provides protection for the instruments when they are not in use. However, all items can be bought separately if you wish to build up your stock of equipment gradually.

Compasses are now adapted to take the reservoir and nib unit of a technical pen, an invaluable aid in inking line work. For airbrush work, it is very useful to be able to fit the compass arm with a fine blade or cutting point for cutting circles and arcs in masking film: it is difficult to maintain accuracy when cutting curves freehand and this is often crucial to a complex and highly finished airbrush illustration.

Ruling and measuring equipment
The minimum requirements for line and colour work are a clear plastic ruler which enables you to see the underlying work as you draw, and a steel rule which also acts as a straight-edge for cutting. The edges of a ruler used for ruling with a technical pen should be bevelled or stepped, to avoid damage to the pen nib and prevent the ink from bleeding underneath the lower edge.

A parallel ruler is essentially two rulers hinged together by an adjustable arm at each end. This item enables you to draw parallel lines at varying widths without moving the ruler. On a larger scale, a T-square extends the capability for parallel ruling. T-squares are available in a range of sizes.

Clear plastic set squares (triangles) are aids for drawing and measuring angles and parallel lines. Useful sizes are 45° and 60°. If your work involves a high degree of accuracy in measurement, an adjustable set square is a good investment, allowing the triangular relationships to be set at different angles.

Include at least one protractor in your drawing equipment for measuring existing angles or setting angles in a drawing. Plastic protractors are available in circular form, showing a full 360°, or semi-circular with a 180° span. These are also marked with basic axis lines.

Templates
A full set of ellipse guides provides templates for ellipses in varying sizes and angles of projection from 5° to 60°, usually increasing in increments of 5°. Again, it is necessary to obtain the best for true accuracy, and this

runs to quite a high expense. To begin with, 30° and 45° projections are the most generally useful, and stock can be built up according to specific requirements. The templates are marked with the axes of the ellipses to allow accurate positioning.

Ellipse templates are continually in use in technical illustration. More specialized templates, such as the spring template for drawing coiled springs, are less often required, but valuable for highly technical line work.

There is a wide range of preformed plastic templates, from basic geometric shapes to graphic symbols and stylized versions of figures and objects. Few of these are appropriate to the majority of technical illustration work. The disadvantage of relying on templates to construct an image is that they tend to dictate the scale of a drawing and the angles of view: the temptation is to build the drawing around the template shapes, rather than use them simply as drawing aids.

Irregular curves, however, commonly crop up in technical illustration, so two or three French curve templates are a necessary addition to the stock of basic equipment, offering a range of broad and tight curves that can be used singly or to form compound shapes.

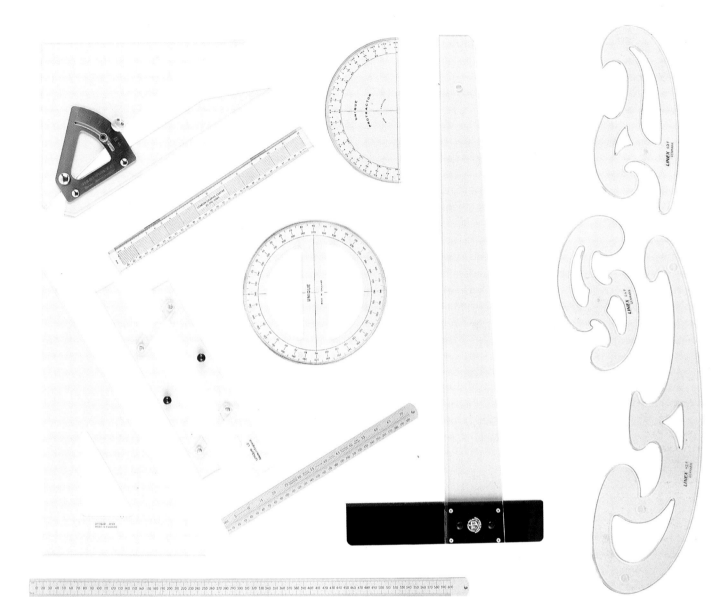

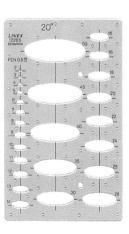

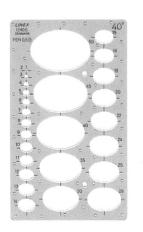

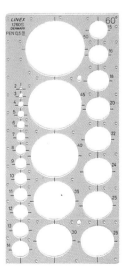

Drawing aids are particularly essential to the technical illustrator because accurate measurement and precise drawing of angles and curves are frequently important elements of the work. Useful basic equipment includes rulers and set squares (triangles) (above), protractors (above left), French curve templates which provide a variety of tight and broad curves (above right), and ellipse templates (right), which are likely to be in constant use.

Printed perspective grids

Grid sheets provide a printed framework for one-, two- and three-point perspective views from normal eye level, ground level and aerial viewpoints. The grids range in complexity from simple block shapes providing an obvious three-dimensional structure to all-over networks of squares and diagonals in diminishing perspective. Measuring systems incorporated in the grids enable the artist to plot receding planes in correct proportion. Grids for plotting orthographic and isometric layouts are also available.

Grids are a valuable aid if you find it difficult to construct or assess a perspective view by eye, but like templates they are best used as a means to an end, not as the definitive means of drawing three-dimensional form.

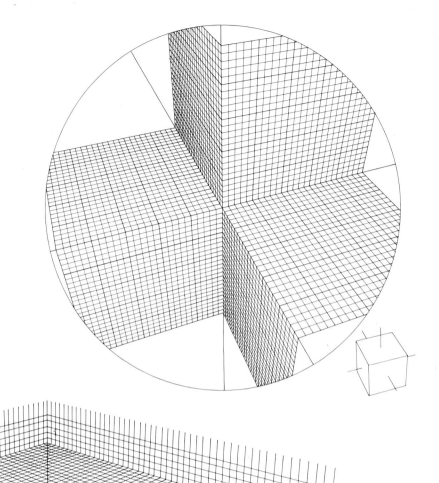

Mechanical tints

This is the term used for pre-printed patterns of fine lines or dots that merge visually into overall areas of tone. The light or dark tone of the tint depends upon the proportion of black and white: for example, fine dots relatively widely spaced read as a light grey tone; heavier and more closely spaced dots create the impression of a dark grey. Literally hundreds of variations are available in the form of self-adhesive film or dry transfer sheets.

The application of mechanical tints is far less laborious than hand-drawn techniques of adding tone to line work. They form a crisp complement to inked lines and add a sense of depth and volume to the surface effect.

Colour tints and flat colours are among the range of these products that is available; these are useful for line and colour work and diagrammatic renderings.

Mechanical tints are rubbed down in place on the artwork. The tint patch can then be lightly trimmed with a scalpel (utility knife) to form clean edges corresponding to the precise outline of the shape.

GENERAL STUDIO EQUIPMENT

The following describes a range of general materials and equipment, from small everyday necessities to the large and costly items which form part of the permanent equipment in a design or illustration studio.

A single glance around the store display of a good art supplier will reveal all sorts of tools, gadgets and materials which are fascinating in their range and inventiveness. New items are appearing all the time, in response to the needs of designers and illustrators, and while it is always possible to improvise for a specific task, it is worth keeping an eye on new materials available. For the beginner, however, the range can be bewildering and an impulse buy may turn out to be an expensive error, so assess your requirements carefully, start with the essentials, and build up your studio stocks gradually.

▶ LAYING MECHANICAL TINTS

1 From the main sheet, cut a section of the tint slightly larger than the area of the artwork to be coloured. Remove the backing paper and rub down the tint patch. Trim the edges, then peel away the excess.

2 Repeat the process to lay an additional tone or colour as required. Trim the edges of the tint patch precisely, following the lines of the orginal drawing, but avoid scoring the surface of the artboard.

3 Make sure that all the tint areas are rubbed down smoothly and that the edges adhere firmly to the artboard surface. Mechanical tints can be used to add even colour or tone to regular or irregular shapes.

Cutting equipment

Probably the most vital item of basic studio equipment for any designer or illustrator is a scalpel (utility knife), a light-handled blade holder which can be fitted with any of a variety of fine, very sharp blades. Angled blades are available of different lengths: some artists favour rounded blades for cutting curves. The scalpel is used for scraping back a surface — lifting an ink line from drafting film, for example — for cutting masking film, trimming mechanical tints, and for an unforeseeable range of day-to-day tasks which may involve scoring, trimming, slicing or scraping. The lightweight quality is important: a heavier craft knife will be needed for cutting thick cardboard or mounts for presentation work, but is too imprecise and clumsy for mask cutting and delicate corrections to surface effects. The fine scalpel blades quickly lose their sharpness, but they are inexpensive and easily replaceable, so keep a good stock of the types you require.

A cutting mat eliminates scratches on the worktable or drawing board and forms a firm, non-slip surface. These resilient rubber mats usually carry a printed grid to help guide accurate cutting of straight lines.

Erasers

There are many different types of eraser to choose from, and this is an item on which people commonly develop a purely personal preference. Plastic erasers for pencil and pastel are efficient and clean. Harder, more abrasive types are preferred for lifting ink or colour. A pencil eraser, with the fine length of rubber composition encased in wood, is helpful for corrections to fine detail, as it can be sharpened to a point like a pencil. This also lifts airbrushed colour quite cleanly, for effects of highlighting and surface texture.

Tapes and adhesives

It is useful to keep a stock of various types of adhesive tape. Low-tack transparent tape is good for assembling tracing paper sketches and for layout work, as it is almost invisible and lifts easily if put down in the wrong position. Masking tape of the opaque, lightly textured kind is strong but also low-tack, allowing easy removal. If paper needs to be stretched for work in watercolour or gouache, you will need a generous supply of gummed brown paper tape.

General-purpose adhesives for studio work include rubber compounds such as Cow gum. This dries slowly enough to allow repositioning and forms a malleable skin when dry, which can be rubbed away where adhesive has squeezed out from underneath the stuck layer. Spray adhesives are the cleanest and most convenient type for working with paper layers. Once you have sprayed the underside of the paper, you can keep sections of an assemblage on a tracing-paper backing until needed, and you can peel them up easily from the artwork and move them to another position without reapplying adhesive. However, there is a hazard involved in their use as the spray particles hang in the air and may be inhaled. Work by an open window or in the vicinity of an air extractor and be sparing with the spray — very little is needed to glue lightweight materials together.

Storage systems

The individual illustrator needs a good-quality, protective portfolio for storing finished illustrations and transporting them when visiting clients. There are many different types, but the most efficient is a sturdy, waterproof-covered portfolio with zipper fastenings on three sides, which includes a spring-lock binder inside for inserting plastic view sheets to protect individual artworks. The plastic sleeves can be obtained singly to accommodate an expanding range of work. This form of combined storage and presentation allows the artist both to preserve and display illustrations in good condition.

Within the studio, a plan chest is the ideal storage system for clean sheets of paper and board and for finished work. This is a solidly built chest of drawers; the drawers are shallow but broadly dimensioned to allow large sheet materials to be stored flat.

Drawing boards

A large sturdy drawing board is a necessary item whatever the type of illustration work you intend to do. For general purposes, buy a board made of dove-tailed blocks rather than a single sheet of wood (this resists warping), with a hardwood or metal strip on one edge to form an accurate

guide for a T-square. Drawing boards are also available with a melamine or other plasticized surface finish, which is non-absorbent and easily cleaned after paints and adhesives have been used.

The ideal studio drawing board is mounted on its own stand, adjustable for height and angle, and fitted with a parallel motion device — moveable straight-edges controlled by a cross-wire or counterweight system, which allow accurate measuring and ruling of horizontal and vertical lines. This is

board, but they combine the supportive qualities of an ordinary drawing board with additional drawing aids, a practical and economical solution.

Visualizers and copying machines

These large and expensive items of equipment are found in design and

illustration studios, providing means of scaling up and printing illustrative images. A visualizer (camera lucida) allows the artist to enlarge and trace off a flat image or take a two-dimensional view of a three-dimensional object. It consists of a moveable copyboard on which the original image is placed. Above this is a bellows fitted with a lens which throws up the image on to a viewing screen when it is lit by the strong light-source flanking the bellows. The bellows is adjusted by turning two handles to scale up and

a major investment for an individual illustrator, and may be out of reach of the beginner or interested amateur who does not make a living from illustration work. However, there are useful table-top boards with angle-adjustment and a moveable straight-edge attached, which are suitable for technical work providing the illustrator is also well equipped with the basic drawing aids required for measuring and ruling. Recently, several manufacturers have also developed ranges of smaller portable drawing boards fitted with vertical and horizontal plastic drafting arms that run on guide rails against a measured scale. Naturally, these do not provide the stability and precision of the free-standing studio drawing focus the image. These components are mounted on a sturdy upright stand, and the viewing screen is usually shaded by a fabric hood, giving the user a clear view of the illuminated image. The artist simply places tracing paper or drafting film over the viewing screen to trace off the image at the required size. In this way, the visualizer saves a lot of time and effort in preparing a drawing.

A good-quality photocopier with facilities for enlargement and reduction is essential in a busy studio for a variety of reasons. But perhaps more pertinent to the work of a technical illustrator is a photo-mechanical transfer (PMT) machine, a process camera which produces photographic copies of an image in black and white on bromide paper. The method of scaling up an image is similar to the principle used in the visualizer, but the PMT machine creates a negative which is passed through a table-top processor to produce the print. This is an extremely useful device for reproducing multiple copies of black line work and for translating a pasted-up image into a single clean presentation. It accurately reproduces line illustration and typographic elements. Some models will reverse the image from left to right, or convert black line on white ground into a reversed white line on black image. There are also machines for colour processing.

Light boxes
A light box consists of a sheet of translucent perspex (plexiglass) or similar durable material mounted over a strong, even light-source. It is extremely useful to the technical illustrator for tracing off stages of a drawing, or for inking on drafting film over an underlay, as the illumination throws up the image very clearly. It is also a means of studying photographic reference supplied in the form of transparencies, when used in conjunction with a viewing glass that magnifies the image. In a large studio there will be one or more light boxes of table size, but there is also a range of small portable versions on the market which are more affordable for the individual illustrator.

A visualizer is an accurate means of enlarging or reducing an image and will normally be found as part of the permanent equipment in a design or illustration studio. The original is placed on the lower tray: when lit the image is thrown upwards onto a glass screen and the size is adjusted by raising or lowering the central bellows. The resulting correctly sized image can then be traced off onto tracing or detail paper laid over the glass screen.

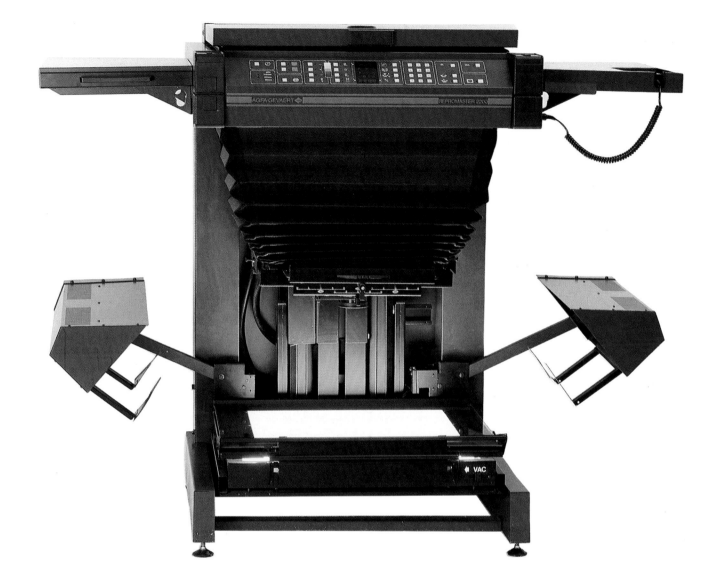

Above: **A PMT (photomechanical transfer) machine is a copy camera particularly useful for reproducing line work and other monochrome images, including topographic material. It also has the facility to enlarge or reduce the image and a clean black-and-white print is produced on bromide paper.**

Left: **A light box is indispensable for a number of uses, such as tracing off an image, checking the fit of an overlay or working from photographic transparencies. This picture shows a small portable model.**

DRAWING METHODS

Skill in drawing is the basic building-block of technical illustration. While highly finished colour renderings may represent the peak of presentation values, they are underpinned by an ability to analyse form and function in technical subjects which is developed by drawing. At its most basic, a drawing process adapts the three-dimensional form of an object to a logical system of two-dimensional representation. An individual drawing is typically selective: the artist extracts certain types of information from the object viewed and translates them into corresponding marks on a flat surface which seem to identify the information efficiently.

There are many reasons for making a drawing and many ways of going about it. A complex technical subject may require several stages of preparation, from the simplest thumbnail sketches to the most detailed and accurate construction drawings providing the basis of a 'realistic' rendering. A drawing is evidence of a thought process, and the more complex the mental analysis has to be, the more varied and extensive the body of drawings that describe it. In the end, all the different aspects of the subject may be brought back together, the artist selecting the essential forms and inserting detail that elaborates the description. This activity produces a single illustrative image displaying all the specific information which the artist considers necessary to the viewer.

THE PURPOSE OF DRAWINGS
Some drawings have a life of their own, while others are strictly a means to an end. The purpose of a drawing may affect the artist's choice of medium and the style of representation. In technical illustration, a drawing can act as a record, an instruction, an explanation, a pattern or model for a different form of presentation, or an enticement to the viewer's interest — and frequently is required to combine two or three of these functions.

Of primary importance to the technical illustrator is the drawing which forms the actual framework of an illustration. This drawing establishes the authenticity of the image, whether the structure is emphasized by inking in line or is overlaid by the surface considerations of full-colour render-

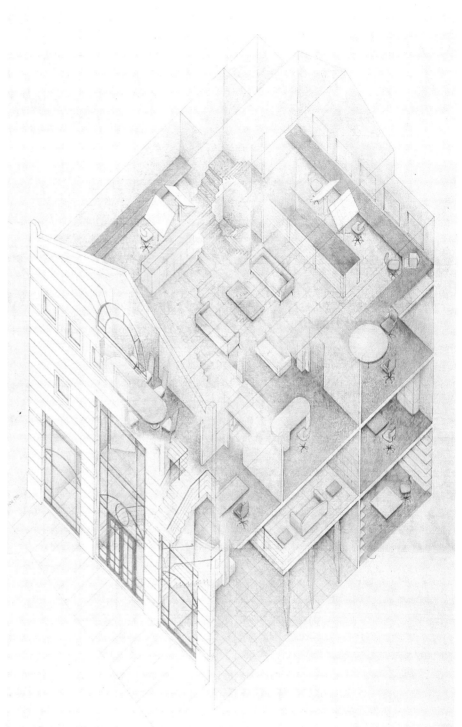

Axonometric drawing
Artist: William White, Pierre d'Avoine Architects
Client: Concord Lighting Ltd
The solid block shape of the formally designed axonometric is subtly elaborated with interior detail and soft colour. The media used are graphite and coloured pencils.

ing. It is both an accurate record of the object and a pattern for the final illustration; it may also be directly instructive. It often follows from a working process which has previously included loosely worked sketches, detailed observational drawings, and possibly a stylishly presented, though not necessarily fully accurate, colour rough. Once the detailed drawing for the illustration is arrived at, it can then be elaborated by selected means to produce the final artwork.

A presentation drawing may involve the artist in visualizing an object as yet not in existence, using plans from designers and engineers to interpret the basic data about the object into an image accessible to a 'lay' audience — that is, to people with an interest in the product but

Above: Automobile sketches
Artist: Alex Padwa
Rapid pen sketches are useful both for working out design ideas and also for becoming more familiar with the general shape and construction of specific technical objects.

Right: Automobile sketch
Artist: Anthony Lo
A feel for line and contour is evident in this ink skekch. The main axes are sketched in to establish the balance of the form.

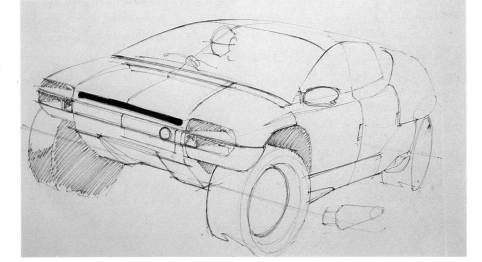

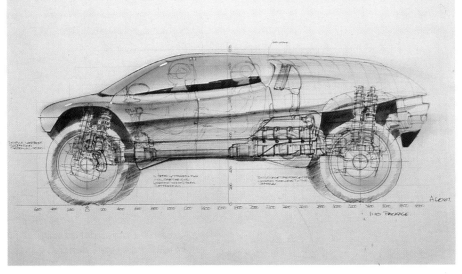

Left: Automobile sketch
Artist: Anthony Lo
Working drawings include specific aspects of technical detail. A brief suggestion of colour enhances the weight and solidity of the design.

Steam train
Artist: Tim Connell, Aircraft
A fully worked pencil drawing has rich qualities of tone and texture. The character of the medium and the techniques of pencil drawing allow highly detailed work to be gradually elaborated with great precision.

themselves lacking the skill to visualize it from existing information. This can be a vital stage in product development and is part of the process of selling an idea. Presentation work does not necessarily involve a high degree of finish in the artwork — as in, for example, a full airbrush rendering — but it has to be both accurately constructed and elegantly descriptive. In recent years, marker renderings have become a favoured technique for presentations, combining the speed and simplicity of drawing mediums with the lavish colour range of more laborious painting techniques.

ACQUIRING DRAWING SKILLS

As a fact of life, the working illustrator is unlikely to find much time to experiment with drawing methods — either

by applying different media to the interpretation of an image, or by extended study of a particular subject through drawing. In practice, illustration work is bound by deadlines and other specific instructions in the brief, and the illustrator must arrive at a sensible estimate of the time and effort which can be put into preparing the image in relation to the time needed actually to produce the finished rendering. Preparatory drawing, in this context, must be seen to make a direct contribution to the work in hand.

Drawing skills, therefore, are ideally investigated and to a large extent mastered before the pressure of day-to-day work places restrictions on the artist's time. For this reason, drawing classes form an important feature of specialized courses in technical illustration, and these may focus on non-technical subjects: anything that has a distinctive form and structure serves as a model for developing analytical and practical drawing techniques. A human figure or complex organic form, such as the layered leaf structure of a cabbage, provides just as valuable an exercise in drawing as, for example, the inner workings of a telephone.

However, the process of learning to draw, which many artists would say is never-ending, can be laborious and even frustrating, as you go over and over the drawing to get it right, constantly referring back to the original model and making adjustments to the

Aircraft foot rudder control pedal
Artist: Lloyd Allen
Portsmouth College of Art, Design and F.E.
These illustrations show three distinct stages, or interpretations, of the same image. The first is the pencil construction drawing worked up by reference to orthographic plans. The second is an ink line rendering, using the convention of two line widths. The third uses colour wash on dyeline paper to render the pedal in solid form.

drawn marks which describe it. To this extent, it is important to be interested in the subject. The discipline of taking a formal interest in a specific object, simply in terms of translating its structure, is one of the elements of drawing skill which only comes with experience. For both the beginner and the student of drawing, an actual fascination with the form and function of the particular subject is a useful spur to concentration.

Seeing and drawing
Typically, people who lack confidence in their drawing abilities focus on the technical skills of mark-making, presuming that it is a matter of learning how to manipulate the pencil, pen or brush more dexterously to form the image in a convincing manner. In fact, the problem usually lies further back, in a failure clearly to see and analyse the real view or object to be represented. In many cases, it is rigorously accurate observation that suggests a method of translating the form into two dimensions. Only then do the technical skills of drawing come into play. An artist's personal style is in one sense merely a means of decoration: six artists' draw-

ings of the same object can each appear very different in purely visual terms, while having a lot in common in their selection of the primary features which establish the object's character. These common elements are the basic physical structure of the form and its volume — angles and curves, projecting and receding planes, and spatial relationships across the overall dimensions of height, width and depth.

Training oneself to see is a matter of practice — practising observation rather than manipulative skill. It is not always necessary to have a pencil in your hand or to move on to the process of translating what you see into drawn marks. Simply looking, and acquiring a discipline of observation, can be done anywhere at any time.

Accuracy in drawing is also a matter of discovering how things really are, rather than viewing them through assumptions based on familiarity. We continually make instinctive assessments of three-dimensional form and scale which enable us to perform everyday activities — to climb a flight of stairs, for example, or grasp a gear lever. To make use of an object, it is not essential to study it in detail. But to

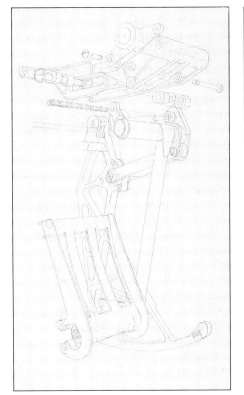

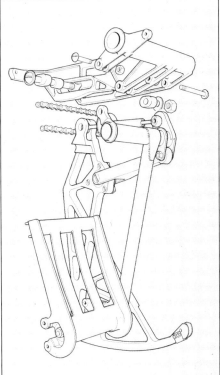

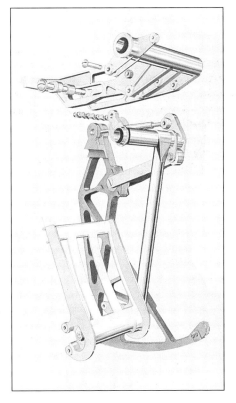

draw an object accurately, you need to abandon the general notion of what it looks like and learn to observe the reality in specific terms — width, height, proportion, angle, colour, texture. These are the 'raw materials' of form which are used to recreate a given view.

OBJECTIVE DRAWING

The discipline of objective drawing involves extended study of the subject to produce a comprehensive drawn image. By the end of the process, this drawing has been thoroughly worked over and revised until the artist feels that every important element of the subject is correctly drawn individually and in relation to the whole, and that at the same time the drawing is properly descriptive of overall form and relevant surface qualities. This is one way in which the student of technical illustration acquires working knowledge of the objects and systems which are the main subjects of this field of work, at the same time developing the manipulative skills of mark-making.

Part of the value of objective drawing is in acquiring a faculty of self-criticism. It is tempting to make do with a drawing that comes 'near enough' to accuracy, but particularly with technical subjects, errors and inconsistencies gradually multiply as the drawing proceeds. If you can see, or sense, that something has gone wrong, there is more to be gained by identifying it and putting it right. The academic tradition of banning erasers from drawing studios has a certain perverse logic: an error which remains visible is a marker from which the correct delineation can be calculated, but an error which has been erased may be repeated.

When drawing a comprehensive view of a single object, it is important to establish an overall view before concentrating on specific areas of detail. By blocking in the main shapes in a way which describes the general forms and their proportions, the artist can ensure that the drawing will fit the required scale and format, and the intricacy of the component parts can be gradually developed. When one

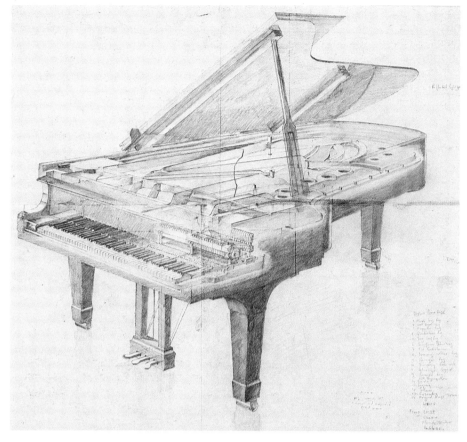

area is worked in great detail without reference to the broader view, it commonly happens that in the gradual extension of the drawing the overall proportions become distorted. To avoid this, it is best to begin work by establishing the general outline and the relative scale of components, before the more complex aspects of form and function are studied more fully.

The term 'drawing' can be loosely interpreted in relation to the media used for objective exercises. Pencil is a remarkably versatile and sympathetic drawing medium, and the one which technical illustrators typically use for general drawing and drawing construction. However, charcoal, pen or brush lines, because of their radically different qualities, encourage the artist to view the subject differently. A drawing might be laid down as blocks of colour or tone, using paints or markers, or collaged from pieces of paper. This experience of handling the materials of illustration and design can be gained only by practical application.

Grand piano
Artist: Steve Latibeaudiere
In an objective drawing of a subject with distinctive form and elements of complex detail, a good sense of the contour and overall volume is required before interior detail can be accurately located. Drawing is an investigatory process providing thorough knowledge of the subject.

SKETCHES AS VISUAL RECORDS

Sketching is a means of rapidly recording information — about objects, systems, components, surface effects — to acquire actual knowledge of three-dimensional forms and develop expertise in translating this into two-dimensional mark-making. This area of drawing skills includes visual 'note-taking' which produces a body of sketches, often including written notes, of different views of an object, details of its components and functions, and possibly the location and surroundings in which it is typically used. The range of detail which can be incorporated in quick sketches also covers descriptive surface elements — colour, patterns of light and shade, and material texture.

This is an opportunity to try out different media to represent varied impressions of the subject. A line sketch in pencil or pen, though rapidly drawn, may be fairly detailed and meticulous; the looser medium of charcoal, by contrast, encourages a broader impression of overall form and basic blocks of light and shade. Coloured pencils, markers and pastels are all easily portable materials for colour notation in sketching, and each contributes its characteristic texture to the drawing. The experience gained in manipulating the materials freely while making sketch drawings may also suggest useful methods of approach to more detailed renderings and finished illustration work.

The ability to sketch quickly and accurately is particularly important in on-site drawing. An illustrator may have restricted access to a busy

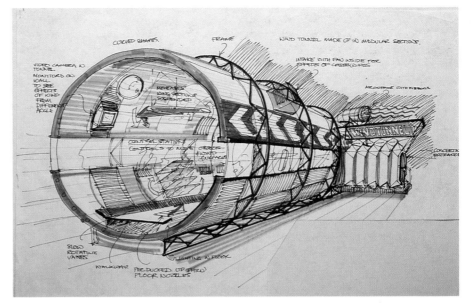

Above: Wind tunnel
Artist: Martin Roche
Studio: Brennan Whalley Ltd
In accumulating information towards a finished illustration, it is helpful to make sketch drawings with colour detail and even written notes which create a body of detailed reference.

Below: Pool table sketches
Artist: John Scorey
Reproduced by kind permission of DPT Snooker Services Ltd
Sketches record many different aspects of the subject, as here the plan and elevations together with notes on detail.

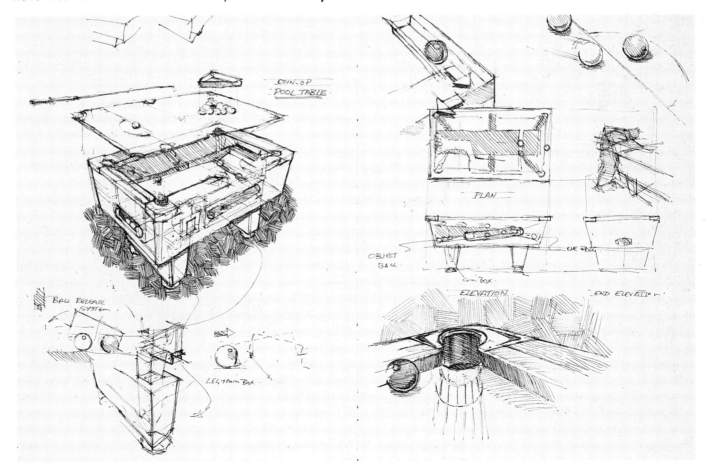

industrial plant or work site, for example, to see a machine at work in its proper context. Such visits can be helpful in deciding on the best view of an object to be taken as the basis of the illustration, although sometimes this aspect of the image may be specified by the client or the designer who assigns the brief.

ROUGHS AND WORKING DRAWINGS

A different approach to drawing comes in plotting the general layout of the image. This process may be the way in which the illustrator begins work on a particular project, or may occur alongside the objective work. Many quite complex images begin life as simple thumbnail sketches, a scribbled outline almost incomprehensible to anyone else, which contains the germ of the idea. The general concept of the illustration includes an overall angle of view; consideration of where cutaway or ghosted sections might be appropriate, or details pulled out from the main image; and the disposition of the parts in an exploded view.

Rough workings of this kind provide the key to the illustrator's approach to gathering information, and as the work progresses, the sketch layouts are developed and fleshed out by additional information. Working drawings are an assembly of detail, though not necessarily as accurate or complex as the final image will be. The amount of stage-by-stage development which goes into preparing an illustration depends upon various factors: the complexity of the subject and the range of possible means of describing it; the time allowed for the whole project (in a professional context, this is also related to the cost-effectiveness of pursuing a broad area of research); and the independence allowed to the illustrator in deciding on the form of presentation.

The illustrator may want to prepare rough layouts that take account of the design context of an illustration. Although there is often a clear demarcation line between design and illustration, so that the illustrator is asked to provide an image for a

predetermined layout, it helps to have a working knowledge of design elements — the format, page grid, typographic style for block text, headings and captions — with which the illustrations are integrated. As part of a design team, the illustrator may also be able to suggest a form of presentation that enhances the interaction of words and images.

Above: British Aerospace Hawk: orthographic presentation
Artist: A. Granger
Redrawn from BAe copyright material
This sheet includes comprehensive information in the form of plan views, front and side elevations and small-scale technical detail.

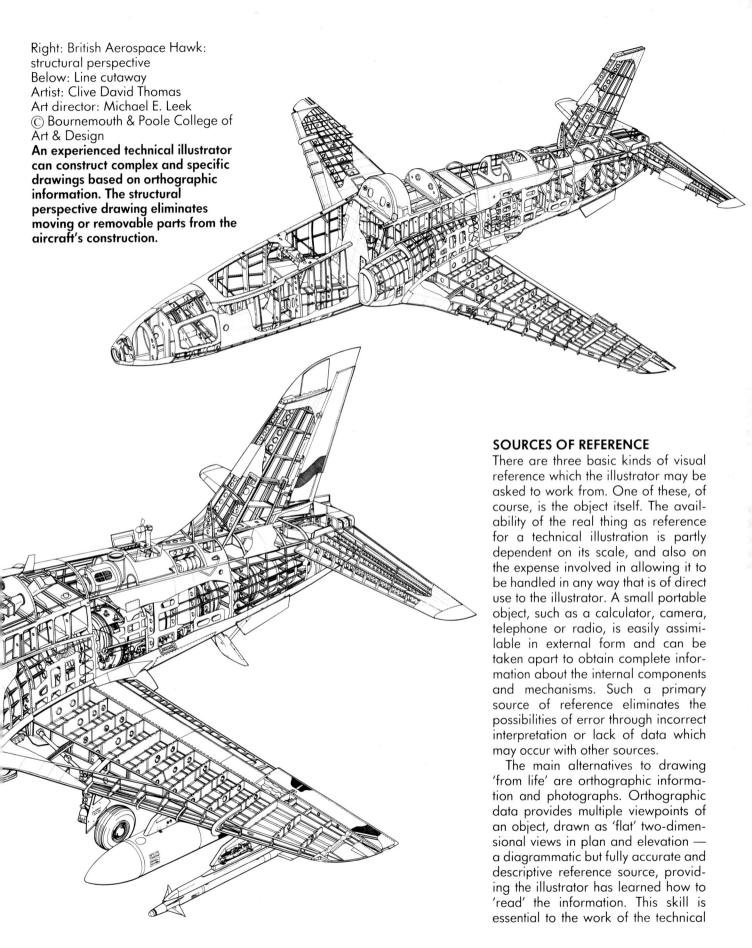

Right: British Aerospace Hawk: structural perspective
Below: Line cutaway
Artist: Clive David Thomas
Art director: Michael E. Leek
© Bournemouth & Poole College of Art & Design
An experienced technical illustrator can construct complex and specific drawings based on orthographic information. The structural perspective drawing eliminates moving or removable parts from the aircraft's construction.

SOURCES OF REFERENCE

There are three basic kinds of visual reference which the illustrator may be asked to work from. One of these, of course, is the object itself. The availability of the real thing as reference for a technical illustration is partly dependent on its scale, and also on the expense involved in allowing it to be handled in any way that is of direct use to the illustrator. A small portable object, such as a calculator, camera, telephone or radio, is easily assimilable in external form and can be taken apart to obtain complete information about the internal components and mechanisms. Such a primary source of reference eliminates the possibilities of error through incorrect interpretation or lack of data which may occur with other sources.

The main alternatives to drawing 'from life' are orthographic information and photographs. Orthographic data provides multiple viewpoints of an object, drawn as 'flat' two-dimensional views in plan and elevation — a diagrammatic but fully accurate and descriptive reference source, providing the illustrator has learned how to 'read' the information. This skill is essential to the work of the technical

illustrator, as so much of the reference provided by engineers and product designers will be given in this form. In some cases it will be the only basis for a projected image of a new product under development.

The basic principles of orthographic views and their conversion to an image with the appearance of three-dimensions are explained on pages 63 and 64. It is also worth noting that blueprints and engineering drawings provided as reference may include other details useful to the illustrator for descriptive reference: simple notation of the materials used to manufacture the component parts of the object, for example, provides the artist with cues for visualizing details of colour and surface texture.

Photographic reference

Photographs are an immensely valuable source of reference, providing comprehensive single views of complex objects and close-up images of individual parts or component assemblies, as well as details of texture and surface finish. Working illustrators use photographs both as the main reference for a selected image and to supplement notes, drawings and orthographic information.

However, it is worthwhile repeating the standard warning at the outset: a photograph tells only one story, and not necessarily a true one. The camera does lie, with extraordinary competence. It is designed, after all, to render three dimensions as a flat pattern of light and shade, and in

Left: Signal box
Right: Photographic reference for signal box project
Artist: Robert Walster

As reference for this cutaway view of an old-style railway signal box, the artist took a number of photographs showing full views and individual components. Photographs are invaluable information when it is not convenient to spend a long time on site, and as a general aid to memory. They also help to establish the best viewpoint for the illustration.

making this transition it both selects from the available information and scatters false clues. A photograph may give a distorted impression of depth and perspective, where the human eye would correct the image by reference to other information stored within the brain. Contrasts of tone can also be flattened or exaggerated by the camera's eye.

This is said merely to emphasize the importance of using photography wisely, since it is obviously a medium of great importance to any visual artist. Photographs can supply information about objects which, for a variety of reasons, are not continuously accessible; they are a quick way of recording overall impressions in a situation unsympathetic to long-term observation; they can supply import-

ant details of an object from different angles and distances in a fraction of the time it would take to sketch the same views. But it is as well to appreciate that the professional illustrator is competing with photography very directly, especially in editorial and advertising work. Photographs provide a good basis for illustration work, but the illustrator is paid to interpret and communicate information in a considered style, not merely to copy a form.

It is useful to build up a photographic library of images culled from magazines, newspapers, and books on specialist subjects. In this way you will have to hand references relating to familiar objects, as well as a range of images outside your personal experience. It is not always possible,

though generally desirable, to take your own sequence of photographs for reference on specific projects. An efficient and versatile 35mm camera is a considerable asset, enabling you to supply yourself with black-and-white prints, colour prints or transparencies. It is not necessary to buy a highly expensive camera, since the photo-

Photographic reference for rocking horse project
Artist: Robert Walster
This composite photographic image reflects the construction of the rocking horse in that different parts of the horse are fashioned from separate wood blocks which are then assembled, giving a good distribution of colour and wood grain pattern.

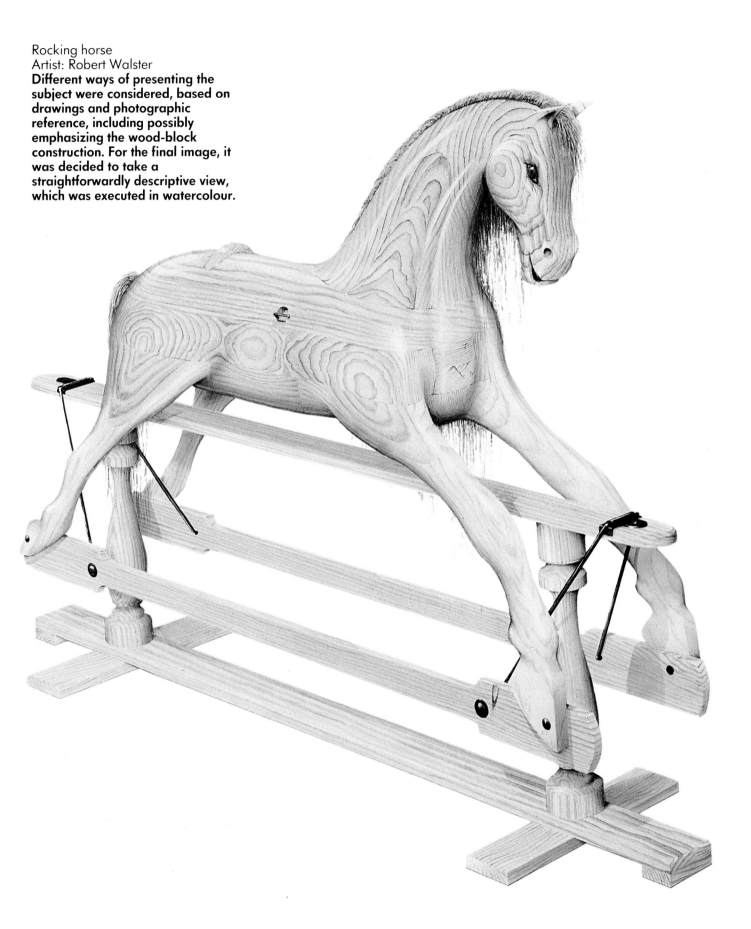

Rocking horse
Artist: Robert Walster
Different ways of presenting the subject were considered, based on drawings and photographic reference, including possibly emphasizing the wood-block construction. For the final image, it was decided to take a straightforwardly descriptive view, which was executed in watercolour.

graphs themselves are a means to an end, not the final product. If you have access to a photographic enlarger, there is great satisfaction to be gained from producing your own prints as required; otherwise, you will need to find a local photographic service offering rapid turnaround on film development and printing.

Polaroid prints are quick and easy to make, and useful as a form of visual note-taking, but are not of sufficient accuracy to be relied upon as a sole reference source.

Using photographs

Given that photographic reference may be only part of the package, there are various ways of putting it to use. The simplest interpretation is a single view, traced and scaled up as a basic drawing for line or colour work. This is a method quite well suited to an individual large object, such as a car or an industrial or agricultural machine. The object is photographed from the required angle and a sharp, large print made, in black-and-white or colour. The outlines are traced over in pencil or fibre-tip pen on tracing or detail paper. A grid is then applied to the tracing and used as the guide for scaling up the line drawing on a proportionate larger grid.

Such an outline trace can also be used as the base drawing in which details of interior structure are positioned for a cutaway or ghosted image. Detailed photographs of individual components, shot from an appropriate angle, provide information for this further elaboration of the image. Separate tracings of part details may be used as underlays to trace off on to the master drawing, especially where scaling up is required, or these details may be worked directly from the photographs.

As an exercise in visual analysis, it is worth taking a series of tracings from a single photograph identifying different aspects of the subject: for example, line or tone analyses explaining spatial relationships; simple tonal blocks showing general distribution of light, medium and dark tones; or areas of the object having distinctive qualities of pattern or texture. Whether or not such drawings act as a basis for finished artwork, they train the eye and brain to separate and select from the various aspects of a subject. Systematic breakdown of the components of an image can also be of direct use in developing a schematic form of rendering.

The illustrator can use a sequence of photographs as supplementary reference for a freehand or constructed drawing. If a general impression of the object is the main purpose of the illustration, rather than a realistic representation, photographs can also be used to form a montage expressing the overall image and relationship of parts. This provides a complex basic structure for interpretation by drawing and colour rendering.

A colour transparency is sometimes used as the basis of a line drawing by projecting it on to a piece of paper taped to a wall. This can be a rather clumsy procedure, as your hand and arm, and sometimes whole body, inevitably block the light from the projector as you work across the image. However, if you can establish an efficient working method, this technique is especially useful as a way of scaling up to a specified size, simply by positioning the projector at a suitable distance from the wall. Distortion is caused if the transparency holder and lens of the projector are not precisely parallel to the drawing surface, so make sure that the image is thrown squarely on to the picture plane before starting to draw.

Various methods can be used for scaling up photographic material mechanically. Some of these, such as visualizers and PMT machines, will be available only to illustrators working in association with a design studio, while others involve equipment within the means of the individual illustrator and are worth buying if working from photographs is common practice.

PREPARATION OF A FINAL IMAGE

There are various perspective and grid frameworks (see page 36) which can be used to systematize the structure of a representation, but many technical illustrators work freehand both when initially plotting the overall scale and spatial relationships in a drawing, and in subsequently working over individual sections of the image to elaborate the detail. Alternatively, methods of drawing construction may be used to establish an overall view which is then developed by freehand drawing.

The various drawing methods make

SCALING UP

1 Construct an accurate grid of small squares on tracing paper over the original reference.

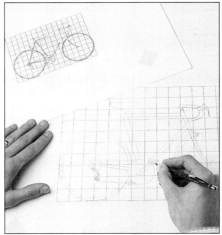

2 Draw up a corresponding grid at the required scale and copy the outlines square by square.

Motorcross crash helmet
Artist: Sean Wilkinson
Drawing with colour is a valuable exercise in both visual analysis and drawin technique. Here a white line is used to ghost in the shapes of the visor and chin strap over the colour work describing the interior texture and helmet casing. Coloured pencil is a quick and clean medium for objective drawing: alternatively, pastels or felt-tip pens and markers can be used, or a combination of media.

use of particular reference points, such as the main axes through the object from a given view, linear cross-sections and definitive changes of plane. Freehand drawing may involve developing a grid system which provides the framework for positioning detailed elements within the overall mass of the object. It differs from working to a pre-drawn grid or strict perspective system, however, in that the artist has more scope to adjust the relationships of different parts of the drawing to ensure that the view of the object appears correct and convincing to the viewer. A drawing which is confined to a pre-drawn grid may

have geometric logic, but does not always correspond to the viewer's visual experience of the object.

From the available reference — the object itself, plans, photographs, sketches — the illustrator acquires a clear understanding of the basic principles and specific components of the individual item, be it large or small. Gradually the picture is built up from a general framework to a recognizable representation, stage by stage. The amount of detail contained in a drawing depends upon the final form of presentation. A drawing which will be inked in line, for example, may include quite intricate

PUSHING THROUGH

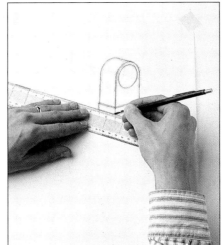

1 Trace off the original drawing, turn the tracing paper and shade across the lines with a 2H pencil.

2 Tape the trace to clean paper or artboard and go over the outlines again with a 4H pencil.

3 Work lightly to transfer graphite to the lower surface. Check the accuracy before removing the trace.

rendering of tiny components or of surface texture, whereas if the drawing is destined to provide a basis for airbrush work, those details which are best described by particular techniques of airbrushing or hand-painting may not feature on the initial drawing.

A drawing which has been prepared for translation into line work or colour rendering typically includes many construction lines which are extraneous by the time the final form has been fully defined. Some of these may be lost along the way, whenever a fresh trace is made using the previous stage of the preparation as an underlay; the underlay is set aside and the drawing proceeds from the current stage. However, the finished image may appear as a complex network of lines from which the selected information emerges. These stages, too, may have been worked out on tracing, detail or layout paper, so the last task is to transfer the image to the required quality of paper or board on which the finished artwork will be completed. The exception to this rule is in the case of line work which is to be inked on drafting film or line-quality paper, when the clean sheet of film or paper can be laid over the drawing on a light box and the relevant line detail traced off.

TRANSFERRING A DRAWING

The traditional method of transferring the outlines of a complex drawing to an alternative surface is a technique known as 'pushing through'. This is basically similar to the commonly used tracing method. The back of the paper is covered with a light film of graphite from a pencil, or may be rubbed over with non-repro blue pencil; this is then placed down on the clean paper or board and secured at the edges with a low-tack adhesive tape. A hard pencil or stylus is used to go over the lines of the original drawing, applying enough pressure to transfer faint lines of graphite or blue lead from the back of the drawing onto the surface of the underlying support.

Sometimes in pushing through, the degree of detail in a complex drawing is partially lost or the line quality may suffer from the practice of tracing over the original. It may be necessary to reconstruct certain areas to re-establish the crispness of the image. This can be done by working in non-repro blue pencil directly on the support — for example, laying in the main axes of a component and using ellipse templates to sharpen up the linear qualities.

An alternative method of transferring a drawing involves the use of transfer, or 'tracing down', paper.

This is a fine paper film coated on one side with a thin layer of graphite (black) or chalk (available in several colours). This takes the place of the colour rubbed on the back of the drawing in pushing through: the transfer paper is simply inserted between the drawing and the clean surface on to which the lines are to be transferred, and is traced over in the same way to leave a faint underlying image.

The advantage of following this sequence is that the original drawing is preserved, in case any problem arises with the final rendering and the work has to begin again. The push-through or transfer leaves only a faint line on the final surface which is eventually covered by ink or paint, or can be erased after inking. If non-repro blue is used in work for reproduction, there is no need to erase any construction lines still visible in the completed illustration.

For airbrush work there is another practical method of achieving this stage of preparation: the drawing can be reproduced photographically as a grey line on bromide paper. A matt surface finish on photographic paper is perfectly suitable for application of airbrushed colour, and this method eliminates the labour of going over the drawing once again.

PRINCIPLES OF PERSPECTIVE

Perspective systems are graphic methods of creating an illusion of spatial depth corresponding to the way in which we perceive three-dimensional forms in space. The principles of perspective assist the artist in translating three-dimensional views accurately into two dimensions; they offer a means of understanding spatial relationships in terms of lines on a flat surface. However, perspectives are theoretical and in drawing act as guidelines for the artist. No system should be allowed to overrule the artist's own sense of the visual correctness of a representation. It is common practice to assess the development of an image by eye at every stage, even if a strictly construc-tional method of drawing is being employed.

The conventions of perspective drawing are geared to architectural scale and embrace a range of views usually outside the requirements of technical illustration. A dramatically foreshortened image, or a bird's eye or worm's eye view of the subject tends to draw out a subjective, even emotional response from the viewer which is inappropriate to the main requirements of technical illustration — to describe actual form or instruct the viewer on the mode of operation of a technological object. The subjects in this field of illustration also frequently limit the degree of perspective construction, since the optical illusions which such systems represent increase with distance. There is very little perspective effect in a straight-forward view of an everyday domestic object, and it is not much increased in something the size of a large haulage truck. Exaggerated perspectives simply appear incorrect at this scale. However, certain visual impressions do occur which, properly reproduced, can authenticate the rendering of three-dimensional forms in drawing, and to this extent a work-ing knowledge of perspective theory is indispensable.

THE VOCABULARY OF PERSPECTIVE

Perspective is a visual device and must be assimilated in practical terms by studying examples of perspective construction. The diagrams in these pages explain the visual effects of the different systems of linear perspective and the stages by which they are achieved. Certain terms are used con-sistently to identify different elements of construction, as defined below.

Convergence

This is the basic principle of perspective which states that parallel lines receding from the viewer appear to converge. Objects positioned along these lines diminish in size with their distance from the viewer. Perspective drawing involves plotting the degrees of convergence and diminishment in relation to the viewer's location, the positioning of objects viewed, and the farthest point of sight, the point of convergence.

To judge a perspective view, the artist imagines a picture plane in front of the object on which the image is cast. The sight line from the artist's eye meets the picture plane at right angles and establishes the level of the horizon line. The object should be within a 60° cone of vision to avoid distortion. It is located on the ground plane in relation to the fixed station point (the viewer).

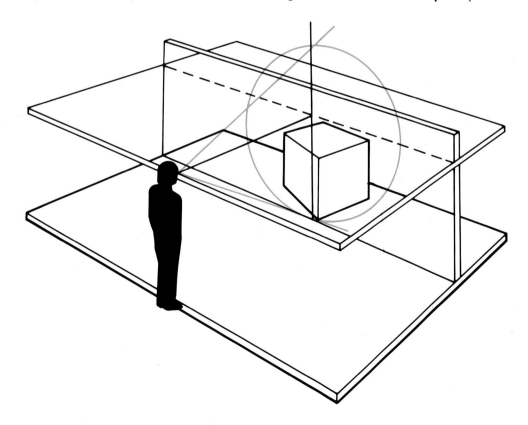

Station point

This is the fixed point from which an object is viewed, essentially the eye of the observer. Also called the viewing point or eye position, the station point is one of the basic fixed reference points from which the perspective construction is calculated. It may be fixed at any distance from the object, and may provide a level, high or low viewpoint on the object.

Centre of vision

The direct line of sight from the station point to the central focus of interest on the object viewed is known as the centre of vision.

Cone of vision

Normal sight includes a range of 180° or more, but some of the information within this range is taken in peripherally. In perspective drawing, a cone of vision is established with a range of 60° or less, since this is the maximum range which can be seen in focus from a fixed viewpoint. The cone represents lines of sight radiating from the eye: its apex is at the station point. The centre of vision is the central axis of the cone of vision.

In establishing the station point for a particular perspective view, the cone of vision should encompass the outer boundaries of the object viewed. Any part which falls outside the 60° cone of vision becomes distorted in the drawing.

Ground plane

This is a horizontal plane on which the viewer or object is standing. The viewer's eye level, or the station point, has a fixed relationship to the ground plane.

Picture plane

This is an imaginary plane on which the image of the object is projected. It is commonly compared to a sheet of glass placed somewhere between the viewer and the object. In ordinary perspectives it is a vertical plane, but it can also be behind the object or tilted obliquely away from or towards the viewer, creating different types of perspective view. The closer to the station point the picture plane is located, the larger the image of the object.

Ground line

Assuming that the ground plane is horizontal and the picture plane vertical, there is inevitably a line of intersection where they meet and this is called the ground line.

Horizon line

The line on which the points of convergence occur corresponds to the eye level of the viewer and is known as the horizon line. It lies on the picture plane parallel to the ground line. The position of the horizon line is selected by the artist according to the viewpoint required. When a low horizon line (close to the ground line) is established, the perspective drawing creates an illusion of looking up at the object from a low level. If the horizon line is high, the view is downward-looking and more of the area of the ground plane is seen.

Vanishing point

The point of convergence of two receding horizontal lines occurs on the horizon line, and this is called the vanishing point. Commonly used perspective systems work around one or two vanishing points. Three-point perspective introduces an additional vanishing point which lies above or below the horizon line, producing a more complex overall view.

Separate sets of vanishing points may be located within a single image to accommodate grouped elements seen at varying angles.

Foreshortening

A horizontal plane seen in plan view, that is, from directly above or below, appears in its true dimensions. When a viewpoint is taken looking across the plane from a position closer to its own level, the apparent depth of the receding plane is progressively less than its actual measurement. The same is true of a vertical plane angled away from the viewer. This effect is known as foreshortening and is the reason why, in a perspective drawing, unlike an isometric or axonometric view, a scale of true measurement cannot be directly applied other than to lines parallel to the picture plane.

The various perspective elements are co-ordinated to produce a logically developed view from a given point and in a given relation to the object viewed. Basic decisions must be made by the artist which subsequently dictate the logic of the constructed image: for example, the position of the station point establishes the size of the depicted object and the angle from which it is viewed; the location of the picture plane in relation to the

In one point perspective, the convergence of lines travelling away from the viewer occurs at a single point on the horizon line. The checkerboard on the left represents a ground plan of cubic blocks on a grid and this is shown three-dimensionally on the right in the corresponding one-point perspective view.

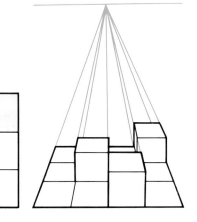

station point separately affects the size of the image; the level of the horizon line establishes the apparent angle of view in relation to a vertical plane.

PERSPECTIVE CONSTRUCTIONS

The effect of a perspective construction depends upon the number of vanishing points employed. To simplify the basic definitions, they are represented here in terms of describing a simple box-like form. Again, it is by following the diagrams that you can understand the actual result of a particular perspective view.

One-point perspective

In this system only one vanishing point is fixed on the horizon line. The centre of vision is at right angles to one side of the object; this side is parallel to the picture plane. Where other plane surfaces of the box are visible at the same time (from left, right, above or below), these planes are receding from the viewer and the lines describing them must converge at the vanishing point already fixed, whatever the angle at which the box is seen. In one-point perspective, lines seen as vertical by the viewer are represented as vertical in the drawing, since they fall parallel to the picture plane.

One-point perspective tends to produce a simplified graphic interpretation of form. It is usefully employed for diagrammatic schemes which include multiples of the same form, since this is easier on the eye than a more complex perspective rendering.

Two-point perspective

Taking again the example of a solid box, in two-point perspective it can be represented with three visible planes, all angled away from the viewer. Two vanishing points are needed to accommodate the convergence of lines from the three planes. Vertical elements remain vertical, as in one-point perspective.

Two-point perspective is perhaps the most commonly used system since it provides a clear view of solid objects and interior structures in an easily assimilable form.

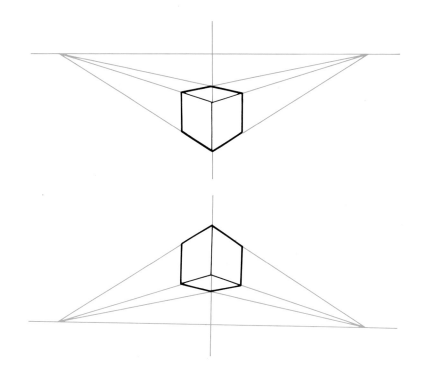

Above: **In two-point perspective, two vanishing points are used to, in effect, govern opposite sides of the object. The location of the object in relation to the horizon line in a perspective view indicates the viewpoint—level, low or high angle. This can be seen from these two examples of the basic cube, one seen from above where receding lines travel up to the horizon, one from below with the lines converging downwards.**

Three-point perspective

In this system, two vanishing points are located on the horizon, governing horizontal convergence, but a third is established above or below the horizon line to allow convergence of normally vertical planes, giving an inclined view of the object. No lines fall parallel to the picture plane.

Using the example of the basic box, the impression is of looking up or

Right: **In three-point perspective each of the visible planes of the basic cube forms an eventual convergence, necessitating a third vanishing point which relates to lines describing vertical planes. This corresponds closely to the viewer's optical experience.**

Below: **Certain constructions require additional vanishing points to regulate the separate convergences of lines describing multiple planes. This might be in the form of a fan-like or rotor-blade arrangement (bottom) and also occurs in the facets of a non-symmetrical angled solid (below).**

down the vertical faces, depending on whether the third vanishing point is above or below the horizon. The planes seem to slope away, a perspective device often employed to render views of tall buidings. With a controlled degree of recession in the third direction, it also creates a convincing impression of a simple, large-

scale object, but the more exaggerated forms of three-point perspective inevitably suggest architectural scale and have a dramatic effect.

Multiple vanishing points
Additional vanishing points may be included on a perspective grid to accommodate changes of plane within a complex form. This occurs when some part of the object is seen at an angle varied from the main view because of its own relation to the object as a whole, not because of a change in the viewer's position. Examples are shown of simple constructions which illustrate this point.

Height lines
A height line is a vertical line on the picture plane, on which scale measurements can be plotted for vertical heights within the perspective view. This means that actual measurements provided with orthographic information (see page 63) can be accurately transferred to the perspective image.

A height line for a specific object is found by projecting from a plan view to locate its position on the picture plane. To simplify the scaling of separate components in a perspective view, height lines can be established for each element of different height as shown.

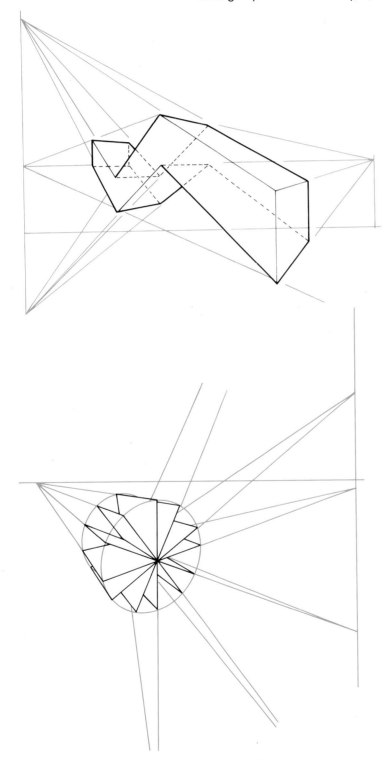

Right above: **Basic perspectives can be constructed without a ground plan, but this is needed to introduce scale measurements based on real dimensions. Extending one side of the object back to the ground line in plan locates the height line on which vertical scale is established.**

Right below: **When working with scale measurements in a composite object, the ground plan can be used in the same way to locate vertical height lines relating to each component element. The correct relationships in perspective are achieved by translating the dimensions of each section into the scale of the drawing.**

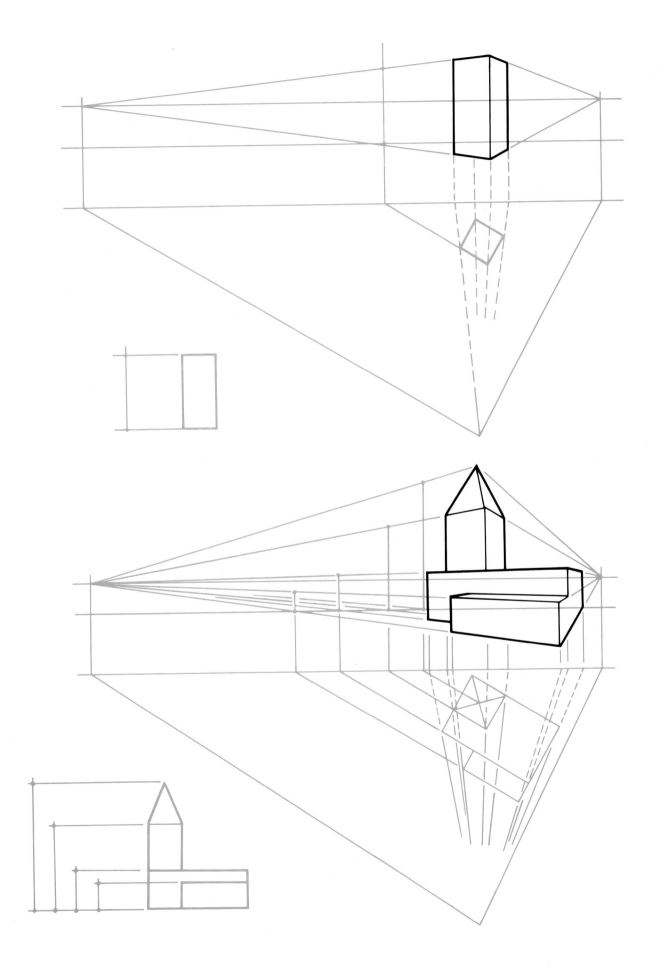

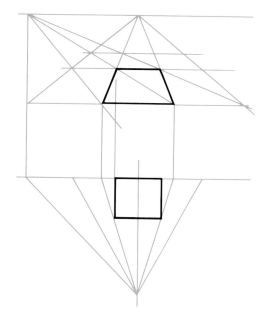

Measuring point perspective

The scale of diminishment of receding elements — the apparent change of scale in uniform elements plotted along the lines of convergence in a perspective view — can be described in accurate relation by establishing a measuring point within the drawing. The measuring point is located by extending the diagonals of a rectilinear form to meet the horizon line at a point separate from the established vanishing points, or at a point on a vertical line passing through an existing vanishing point. A measuring point can be used to project the construction of a series of identical forms with the correct effect of perspective diminishment as seen from the fixed station point.

Above: **To plot the perspective effect of apparently diminishing scale in receding objects, a measuring line is drawn related to the ground plan and perspective view of a single element. This locates a vanishing point for diagonals which creates proportional relationships for similar elements.**

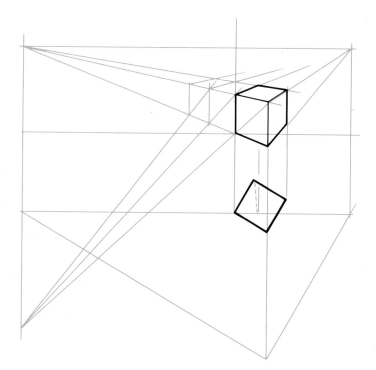

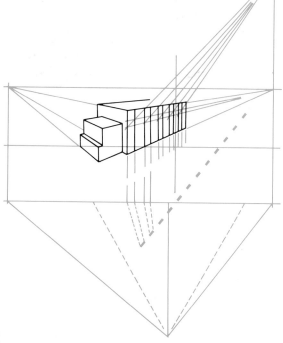

Above and left: **The measuring-point principle shown in the checkerboard arrangement (top) is adapted to cubic forms by including diagonal lines through the vertical planes of the cubes. Similarly, it applies to other constructions capable of being divided on a rectangular framework (left) or irregular forms which can be accommodated by a grid system.**

DRAWING CONSTRUCTION METHODS

Perspective theory explains the basic spatial relationships utilized in developing graphic representation of three-dimensional objects. The practical application of drawing constructions in technical illustration includes various different methods of plotting three-dimensional information to produce a flat rendering which either appears correct or is accurately measured and reproduced, depending upon the degree of technicality required.

While it is not possible here to present every aspect of drawing construction which the technical illustrator may encounter or need to know in studio practice, the following section introduces commonly applied practical skills relating to interpretation of particular forms. These can be used as significant clues towards devising graphic equivalents for related forms and compound objects.

ORTHOGRAPHIC PROJECTION

This system of representation describes three-dimensional form by a series of diagrammatic views in which the information is 'flattened out'. The essential elements of an orthographic projection are a plan view of the object, a front elevation and an end elevation — that is, direct views of the object (as if at right angles to the line of sight) from three different sides. This is the minimum needed to explain the overall structure and volume unambiguously. The three basic views may be supplemented by additional views and sections through the object, depending on the complexity of its construction.

Engineering drawings present information orthographically, and are often a major source of reference for the technical illustrator, particularly crucial to presentation work where the object does not yet exist in solid form and the illustrator is required to project a realistic rendering. Orthographic projections can be translated

Above: **Orthographic projection presents an object as a series of flat images—plan and elevations. Unbroken lines represent a visible outline, dotted lines show changes of plane not visible from the given viewpoint, an example here being the hole through the bracket base.**

into three-dimensional renderings through application of perspective systems or by construction of a basic grid framework (see pages 59 and 65). There are two main systems of laying out orthographic information which may be encountered in studio practice: first-angle projection and third-angle projection.

First-angle projection

In this arrangement of the plan and elevations, the front elevation is presented with the plan view directly below, as if itemizing what is seen when looking up from below the

object, but the end elevation is projected at the opposite end of the front elevation to that which is its actual location.

Third-angle projection

In this view the front elevation lies below the plan, following a logical progression as if looking down directly on the object with the plan view in direct sight and the front elevation below. The end elevation is projected out in line with the normally adjacent side of the front elevation.

Below: **There are conventional ways of presenting orthographic projections. Two standard formats are first angle projection (left) and third angle projection (right). These show basically the same information but with the elements differently ordered in relation to each other.**

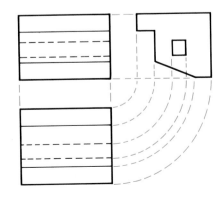

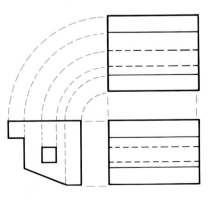

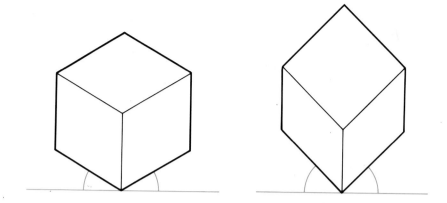

Right: **Isometric (left) and axonometric (right) projections appear to the viewer to be tilted as compared to perspective views. They do not correspond to normal perception of three-dimensional objects.**

The line conventions of orthographic views indicate changes of plane using a solid line for visible detail from a given view, and a broken line for hidden detail. The diagram below shows an example of how these lines relate to the three-dimensional structure of a solid form.

ISOMETRIC AND AXONOMETRIC PROJECTIONS

These are non-perspective means of constructing a three-dimensional view based on orthographic information. An isometric view employs vertical lines and receding axes at 30° angles from the horizontal; all ellipses are 35°. This form of representation appears incorrect three-dimensionally — any horizontal plane appears to rise — because there is no allowance for the principle of convergence as used in perspective systems. The advantage in technical drawing is that actual measurements can be represented on a single scale, as the view is not foreshortened.

In axonometric projection the plan view provided by the orthographic information remains parallel to the picture plane, so true measurements can again be used in scale. Vertical lines remain vertical, but the construction appears tilted forward from the ground plane. Receding axes can be set at any selected angle from the horizontal, but 45° is commonly used.

There are various other methods of constructing graphic equivalents of three-dimensional representation from orthographic information, including oblique, dimetric and trimetric projections. These systems may be of significant interest to the technical illustrator in particular working contexts. It is recommended that readers who wish to pursue the subject further consult specialist publications which describe the full range of drawing projection methods in detail.

Practical construction methods
The exercises in these pages demonstrate methods of using basic construction elements to plot three-dimensional form. The general principle of spatial depth is established by means of a regular cube, which may be used to govern the overall volume of a construction. The ellipse is another construction unit of primary importance in technical illustration, and this is related to both general forms and specific representations in the later exercises in this section.

Below: **To construct a cube receding from the viewer symmetrically at a 45° angle, begin by drawing up the base square in relation to two vanishing points; its front and back corners fall on a vertical line halfway between the vanishing points. Put in the diagonal from left to right. Raise an arc from the right-hand corner with radius equal to the diagonal. Intersect this with a line angled at 45° to the diagonal. From the point of intersection, draw a horizontal line which forms the diagonal of the top face of the cube to establish the height of its outer sides. Working from the vanishing points, draw in the top and side planes.**

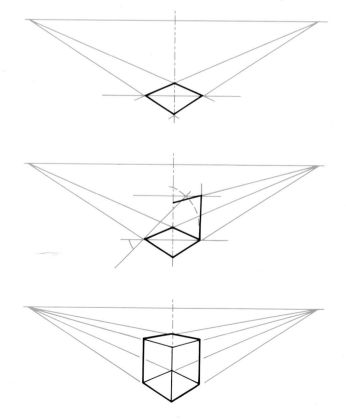

Developing a cubic grid

A cube represents a box with equal sides into which a regular or irregularly shaped object can be fitted, with one or more points on the object's surface touching the sides of the box and thus forming reference points for plotting the overall shape. Other rectangular proportions can be accommodated by regarding the cube as a unit of structure and creating a box which is, say, two cubes wide and three cubes deep. A system of dividing the cube by drawing in the diagonals on each side creates a complex grid with various reference points relating to main axes and planes.

This drawing method can be extended to provide a framework for an object of any given proportions. The cube which forms the starting point for the grid can be created by geometric methods for constructing a cube at a selected angle of view, or by developing a perspective grid around a freehand-drawn cube representing the overall impression of the view required by the illustrator.

Ellipse constructions

An ellipse is a circle in perspective. To consider this simple proposition, imagine holding up a plate in front of you so that you see it as a flat circle. Hold the centre point on each side and tip the plate backwards. The horizontal diameter remains constant, but as the plate tips from the vertical the depth which was represented by the diameter of the plate from top to bottom appears to become progressively less. The circle has become an ellipse; the ellipse narrows with the increasing angle from the vertical.

The lines which connect the widest points of the ellipse — in plan, two diameters of the circle set at right angles to each other — are the axes of the ellipse. The longer is called the major axis, the shorter is the minor axis.

When the ellipse represents a circle lying parallel to the ground plane in a perspective view, the drawn axes appear as a horizontal and a vertical line in the drawing. But when the circular plane recedes at an angle

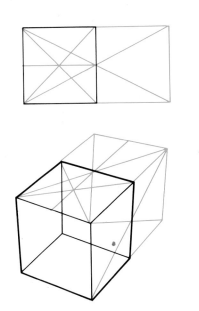

Above: **To extend a cube into a multiple grid, first consider the plan of the cube as a square. Draw the diagonals and find the centre of one side. Extend lines through that mid-point to construct a second square of equal proportion. This configuration can be extended to a perspective view of the cubes by using the diagonals and mid-points of the sides corresponding to the squared plan. Always use diagonals to cross-check the proportions of the cubes as the grid framework is extended.**

Below: **An ellipse is a circle seen in perspective, as demonstrated by these diagrams which illustrate the circle 'in rotation' side to side and top to bottom. The more the flat shape of the circle appears to recede from the viewer, the narrower the ellipse. In the same way, ellipses can be drawn on various axes within the form of a sphere.**

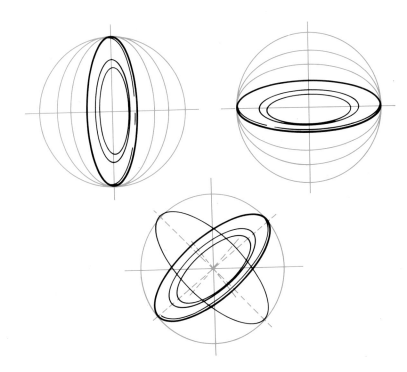

from the viewer, the axes are also angled, although they remain in constant relation to each other, intersecting at 90°.

The ellipse is the key to perspective drawing of forms with a circular cross-section. A cylinder, for example, can be seen diagrammatically as a series of ellipses joined by straight lines through the outer points of the major axes.

Whatever angle the cylinder is viewed from, its central axis passes through the intersections of the ellipse axes. In fact, the central axis of the cylinder corresponds to the minor axis of an ellipse at any given point on the cross-section. This makes it possible to establish the major axis by drawing a line at 90° to it.

The proportions of an ellipse which forms the plane surface on a cylinder can be established by projecting from a plan view of a circle inscribed in a square. The circle touches the mid-point of the square on each side. If the square is turned to become, effectively, the receding plane on one side of a cube, the mid-points can be used as reference to plot the ellipse corresponding to the circle within the square. Given a central axis for the angle of the cylinder, the axes of the ellipse can be found as explained above.

The proportions of the ellipse are not constant along the length of the cylinder because the angle of the circular plane is changing in relation to the distance from the eye of the viewer. Ellipse templates correspond to different angles of view and are measured in degrees. In practical terms, it may be impossible to find an ellipse template to fit precisely a specific element of a drawing, and in this case the nearest size representing the required view should be used. It is difficult to draw an ellipse purely freehand, but templates are not always available or appropriate for all possible drawing problems. If you find it difficult to produce a clean shape working freehand, there are methods of plotting an ellipse based on a circle which can also be used as a guide; these are explained in the illustrations on page 67.

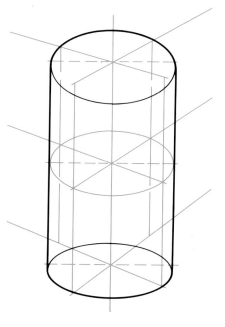

Above: **As seen in plan, a circle representing the section view of a cylinder fits within a square. Projecting this as one side of a cube, the circle becomes an ellipse, still contained within the square now seen at an angle to the viewer. Extending this into a cylinder, the farthest ellipse is seen in a more open view than the nearer one.**

Right: **In this vertical cylinder, upper and lower ellipses are divided as in the circle-in-square plan view (above). Verticals dropped from the points of intersection provide reference points for measuring ellipses within the cylinder.**

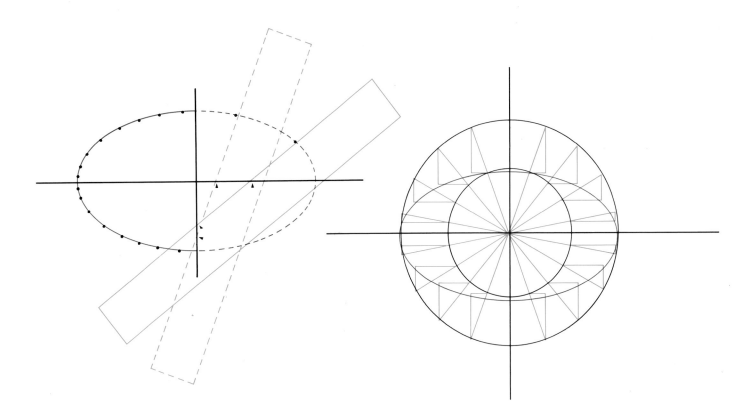

Concentricity

A car tyre is an example of an object which is circular in cross-section and has component parts which are concentric — simply, the hub cap and surrounding tyre share a common centre point at the intersections of their axes. However, the surface of the tyre is not uniform: both the hub cap and the rubber tyre have moulded forms that create a ring pattern of convex and concave circles.

In a perspective view, with the face of the tyre angled away from the viewer, the basic form becomes a pattern of ellipses rather than circles. Drawn from a common centre, this pattern creates a flat, rather unrealistic effect. A more convincing view is produced by shifting the ellipse guides in and out from the centre line to vary the width between the curves, representing the variations of surface level across the form.

This is one of many ways in which the illustrator adjusts the geometry of a rendering to correspond to visual effect. Judging the apparent correctness of a construction by eye is a necessary skill in technical work.

Above left: **This shows the principle of tramelling an ellipse. Draw major and minor axes and mark half the length of each on a strip of paper, measured from a given point. To plot the ellipse, keep the first mark aligned to the minor axis, the second to the major axis, and use the third to plot points forming the ellipse.**

Above right: **An alternative method uses lines projected from the circumferences of concentric circles and meeting at right angles. The points of intersection form the ellipse.**

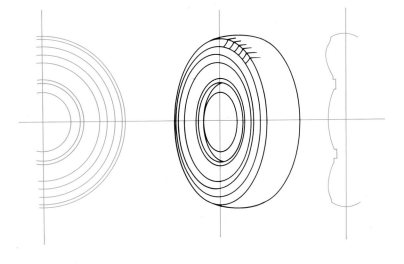

Left: **The axes of concentric ellipses do not precisely align. To reflect this, in drawing a perspective view of an object with concentric structures, as in the rings of a tyre, the ellipses should be shifted slightly in relation to each other to provide a more realistic view. This can usually be done by eye. This drawing of the tyre shows the ring pattern in plan, perspective view, and profile.**

Drawing component elements

There is an immense range of engineering components which form recurring elements in technical illustration. Many of these are very small-scale, and part of the training of a technical illustrator will include familiarization with specific components and methods of drawing them from different angles. Screws, nuts and bolts, springs, brackets, cogs and gearing mechanisms form the mundane but essential content of, for example, line work for technical manuals. In the broader sweep of advertising and editorial illustration, where bold, glossy images are rendered in full colour, these details may be completely lost or irrelevant. However, the general principles of their construction provide useful exercises in mastering particular methods of drawing different forms which may also have application to much larger-scale items.

Through the exercises illustrated here, a representative range of forms is presented. Though far from comprehensive, they provide guidelines to solutions in drawing construction. As can be seen, ellipse templates are invaluable in this type of work: they reduce the illustrator's labour in drawing up components which have linear curves relating to fully or partially circular planes, and they establish a crisp regularity in the image. The exercises can, however, be attempted freehand to gain a practical feeling for the means of construction.

Right top and centre: **A simple screw thread is constructed as a series of partial ellipses. Line and block shading will provide a sense of continuous form. The more open spiralled thread (centre) is based on two concentric cylinders representing the solid core and outer spiral.**

Right below: **The coils of a spring can be drawn using ellipse guides to create the curves over a grid construction which plots the continuation of the coil as, in simple terms, a zigzag line within a rectangular framework based on the ellipse axes.**

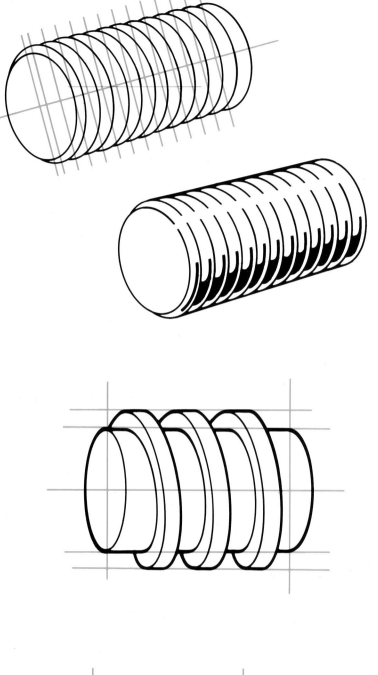

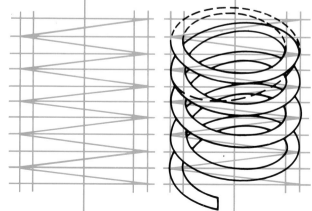

Right: **An alternative method for drawing a spring is to use a spring template which is a guide for the way the coil construction spirals through the elliptical patterns of a simple cylindrical form. The template shape represents the back curve of one turn of the coil passing down into the forward curve of the next.**

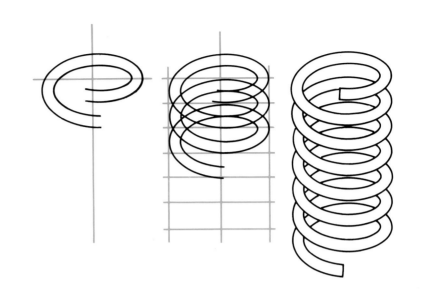

Below: **A complex construction such as a toothed gear wheel can be plotted by projecting from a plan view. This provides guidelines for positioning the regularly spaced elements within the angled view, accommodating the effect of foreshortening which makes the spacing of the gear teeth appear to vary.**

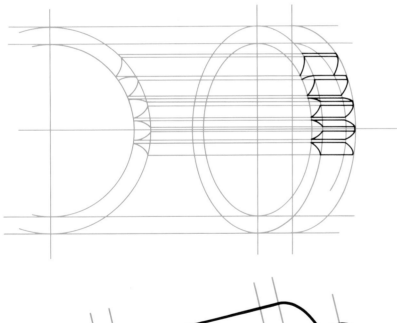

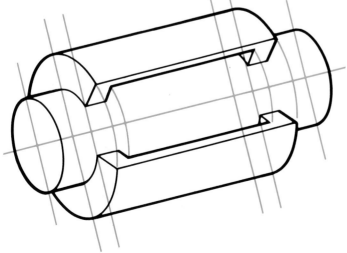

Drawing a section cutaway

The cutaway is an important convention of technical illustration, explained and illustrated in more detail in section 9. The basic convention of drawing a simple section cut into a cylindrical form is demonstrated here to establish some of the general principles and specific techniques of describing a cutaway section.

The 'pie slice' section as illustrated is a simple but very effective way of increasing the range of information about the object which can be presented in a single drawing. A section developed from a centre line provides a balanced and graphic explanation of the interior. It is important to make the central angle of the section large enough to open up the interior view clearly. If there is a central shaft within the cylindrical object, it creates an authentic impression to show its actual contour, developing the cutaway in successive layers leading out towards the exterior surface.

Left: **A simple cutaway view of a cylindrically constructed object shows the wedge-shaped cut towards the axis of the cylinder at an angle of more than 90°.**

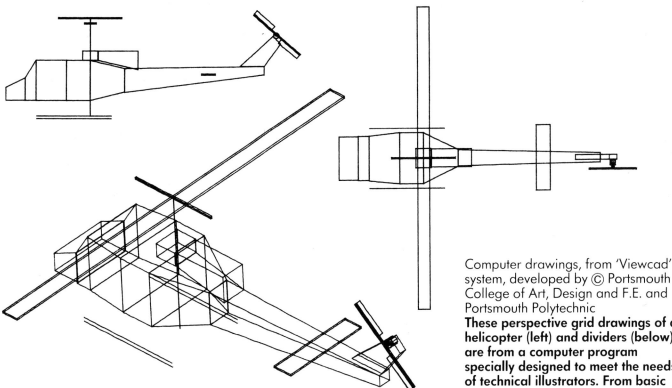

Computer drawings, from 'Viewcad' system, developed by © Portsmouth College of Art, Design and F.E. and Portsmouth Polytechnic

These perspective grid drawings of a helicopter (left) and dividers (below) are from a computer program specially designed to meet the needs of technical illustrators. From basic data, the computer can quickly produce alternative angles of view, enlarged details etc.

COMPUTER DRAWING

Drawing is the area where computer functions have most immediate impact on the current generation of technical illustrators. Graphic delineation of three-dimensional form is readily achieved by this method. The computer can eliminate much of the labour of preparing a drawing by providing an almost instant framework, leaving the illustrator much more free to develop the interpretation of the image in suitable form.

The computer has the significant capability to project, given the essential basic data, the three-dimensional structure of an object and innumerable views of it from different angles. The illustrator can assess these visually on screen and select the required view, which can then be run out as hard copy. This provides both inspirational and practical aid to developing the layout of an illustration, but because the information provided is in schematic form, the artist has open options on subsequent visual style.

At the scale of individual components commonly used in engineering and manufacturing contexts, computer drawing may relieve much of the tedium that a technical illustra-

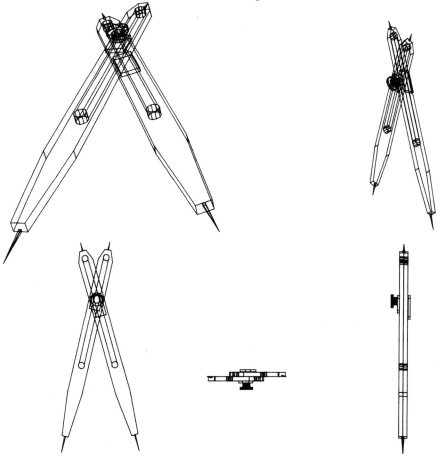

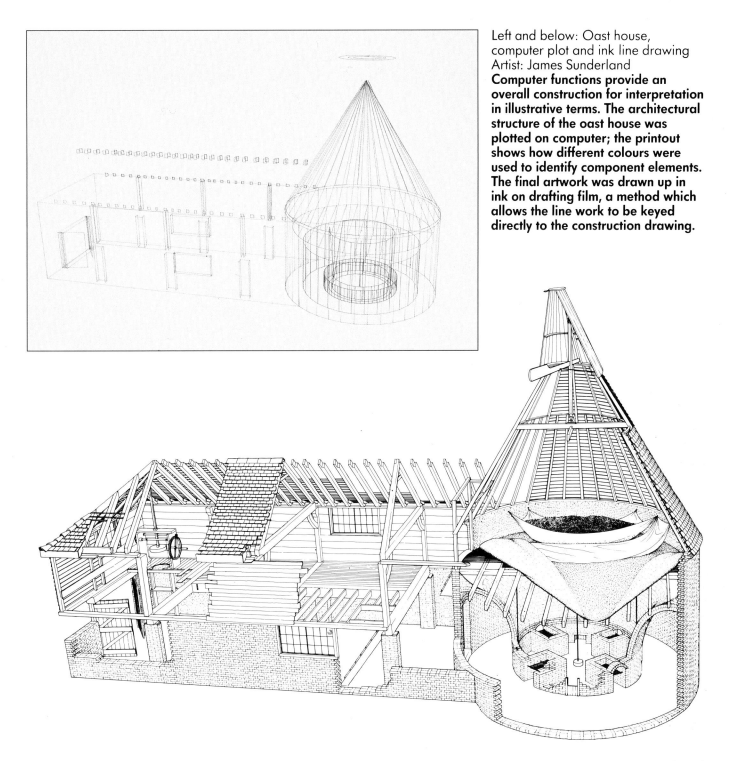

Left and below: Oast house, computer plot and ink line drawing
Artist: James Sunderland
Computer functions provide an overall construction for interpretation in illustrative terms. The architectural structure of the oast house was plotted on computer; the printout shows how different colours were used to identify component elements. The final artwork was drawn up in ink on drafting film, a method which allows the line work to be keyed directly to the construction drawing.

tor's work can involve. Patterns of specified components can be produced in varying scale and viewpoint. These can then simply be used as a master sheet for separate elements of a drawing, which the illustrator can scan to select the appropriate item as required.

The 'building block' approach of computer drawing enables the illustrator to devise an abstract framework or three-dimensional grid by selecting basic geometric shapes for projection in three dimensions. These can then be 'pasted' together on screen to create compound forms. Such an image may be used as the grid for drawing an object in which the main structure demonstrates those simple formal relationships. Computer functions can also be employed in plotting serial relationships in two dimensions as the basis of hand-drawn artwork. Alternatively, the computer image may be elaborated until a recognizable linear rendering is achieved. Graphic facilities commonly now in use can also supply tint areas and textural patterns to be dropped into the line work.

LINE WORK

The crispness of black line on white ground gives a unique clarity to this area of technical illustration, which forms a high proportion of bread-and-butter work in the professional field. Line work may lack the glamour of full colour rendering, but it has a long-established place in technical illustration and the techniques have been developed to a sophisticated level. A well-executed line drawing has considerable aesthetic impact, and it is also a challenge to the illustrator, especially when working with such economical means to form a detailed and explanatory image of a highly complex object.

PREPARATION

The methods of constructing an image, as explained in the previous chapter, are common to line and colour processes. When the illustration has been drawn in pencil in full detail, it can either be transferred to a smooth-surfaced artboard (see page 56), or used as an underlay for inking on drafting film — essentially tracing off the image on the film using a technical pen rather than a pencil. This method saves time on transferring or 'pushing through' the original image. However, many illustrators prefer the surface quality of paper or board, and a smooth paper suitable for line work, such as CS10, can be placed over the original drawing on a light box to be traced off in the same way.

There is always a danger that some detail may be lost in tracing off. After transferring an image by pushing through, you may need to go over some parts of the drawing again. To avoid this, it is sometimes the practice to draw up an image directly on paper or board using non-repro blue pencil, a method which gives immediate accuracy.

Drawing ink does not adhere well to a greasy surface. To remove any greasy marks which may be already on the surface of a line board, it can be prepared before inking by dusting with pounce, a white talcum-like

Left: Beryl A oil platform
Artist: Ron Sandford
Client: Mobil North Sea Ltd
There are obvious restrictions on line work as compared to colour rendering but monochrome images can be equally exciting when variations of texture and tonal contrast are well used for dramatic emphasis.

Right and below: Steam traction engine
Artist: Darren Madgwick
Portsmouth College of Art, Design and F.E.
There is no means of disguise in line work, the original drawing must be an accurate and visually interesting base. These examples show how the complex detail of the pencil construction drawing is translated into a clean, crisp ink line illustration.

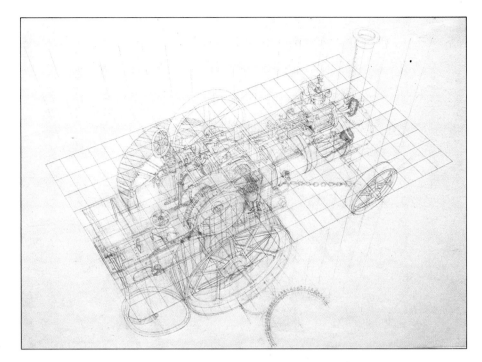

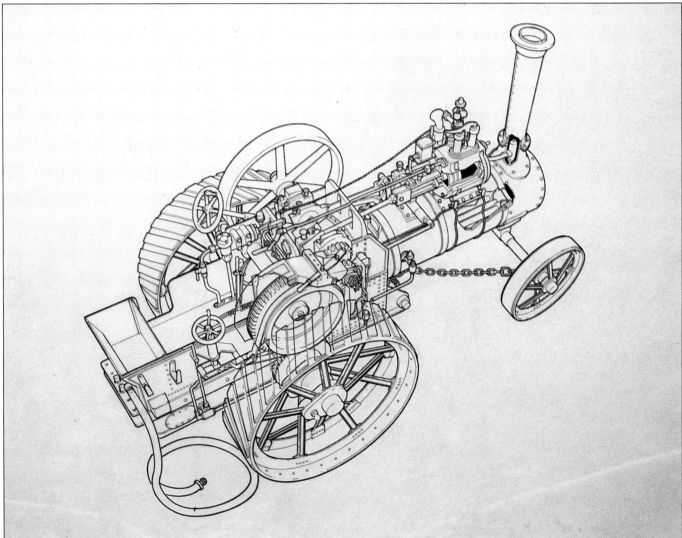

INKING A CURVE

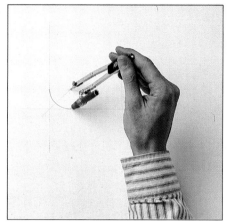

1 Construct the curve accurately in pencil, using a ruler and compass to establish a smooth line.

2 Use a technical pen attachment on the compass to ink the curve evenly in one clean sweep.

3 Extend lines of equal weight from each end of the curve using a straight-edge and technical pen.

powder. Some illustrators shave the surface finely with the edge of a razor blade, which not only degreases but also removes any minor inconsistencies in the board finish.

INKING TECHNIQUE

Since by this stage there should be no remaining areas of ambiguity or unresolved form in the image, the inked lines follow precisely the established structure of the drawing. The next task is to devise a working procedure which reduces the possibility of error or of spoiling the image inadvertently. An important element is protection of the drawing surface throughout the process. To avoid leaving fingerprints or smudging lines already inked, a clean sheet of tracing paper may be used to cover the area beneath your hand and arm as you work; you can cut a window in tracing paper to form a mask surrounding an area of detail.

Some illustrators proceed by starting at the top of the image and working downwards. Alternatively, parts of the image can be sectioned off, so you work up one section in detail before moving on to the next. While working systematically through the image in this way, it is necessary to take an overview from time to time to check on the consistency of the drawing. Variations in line weight are often produced by 'doubling up' the line

rather than using a thicker pen nib, for example, so variations can occur which may go unnoticed unless the overall effect is monitored.

It is usual to ink curving lines before straight lines on broad shapes, especially where they form a link in a specific structure, as it is easier to extend one end of a curve smoothly into a straight line than to join the curve accurately to one or two existing straight lines. In detailed work involving parallel lines and small curves, however, such as a system of fuel lines, more accuracy may be gained

by inking the straight lines first, as the curves are on a tight scale.

When using a plastic ruler or template to guide the technical pen, there is the danger of ink bleeding underneath the guiding edge. This problem is avoided if the lower edge of the template is not held flat on the drawing surface: many templates are made with bevelled or stepped edges for this reason. Otherwise, one template can be placed over another — provided this can be arranged leaving the guiding edge free — to raise the underside of the template in use

INK LINE CORRECTION

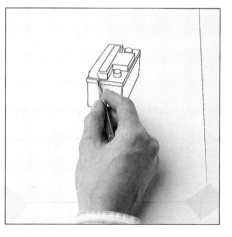

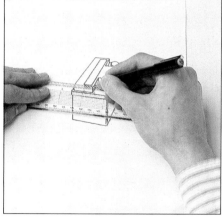

1 When an error occurs in inking, allow the ink to dry and scratch out the line gently with a scalpel blade.

2 When the erasure is complete, use a straight-edge and pen to re-draw the line clearly in the correct position.

from the surface. Make sure that it is not raised so high as to be at an awkward level against the nib unit of the pen, which might cause inaccuracy.

Alternatively, strips of masking tape can be attached to the underside of the template to cushion and raise it. This is a practical solution provided that it does not interfere with your ability to make accurate use of measurements or axis lines marked on the ruler or template. The masking tape rapidly accumulates dirt, such as graphite powder from drawn pencil lines, but can be replaced easily as often as is necessary.

Do not move the guide too quickly after the lines have been inked, and lift it from the surface rather than sliding it away. Hasty movements may cause smudging of the inked lines.

If a line is drawn in error or a bleed occurs destroying the crispness of the line, the ink can be erased using the edge of a scalpel (utility knife) blade to scratch back the dried medium. Alternatively, use a hard ink eraser or specially formulated solvent. The choice is largely a matter of personal preference. If corrections are made by scratching back, take care not to allow the knife blade to dig into the surface. Curved blades are often preferred, as these minimize the risk. All methods of erasing must be used with care to avoid damaging the surface of the artboard or drafting film, especially if you are going to ink a fresh line over the erasure.

LINE WIDTHS

Sometimes the working illustrator will be given a house style to follow or specific instructions from a client on the means of representation, if these have to conform to particular levels of information for the viewer's instruction. At the other end of the scale, in editorial illustration for example, the formal presentation of line work may be left very much to the illustrator's own choices, as long as the finished work does the job of visual explanation required by the client's brief, and is of suitable quality for reproduction.

Conventions in line work have changed in recent years, but one convention still widely used involves two

widths of line defining different aspects of the form. The heavier line is used for outlines. This implies that there is space beyond the boundary represented by that line. It may be simply the space surrounding the object, or a gap between separate but linked components, or an aperture in a plane surface. The lighter of the two line widths represents changes of plane on the surface of an object or component, such as angled edges, areas of concavity or the junction between interlocked components. This line quality may also be used to describe textural elements, such as grilles,

Motorcycle: reflected image
Artist: Roger Savage
Portsmouth College of Art, Design and F.E.
The device of the mirror image presents two directly comparable views of the object. Details of the machinery are given in the main image with ghosted lines showing the overlying position of the motorcycle's bodywork: in the reflected image those ghosted areas are seen in solid form. The overall view was achieved by photographing a plastic model on a mirror. Additional information came from a full-scale machine.

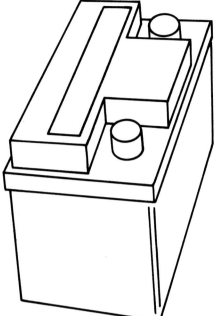

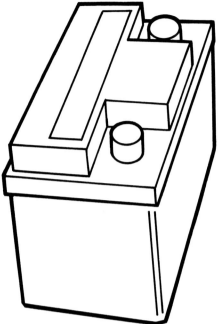

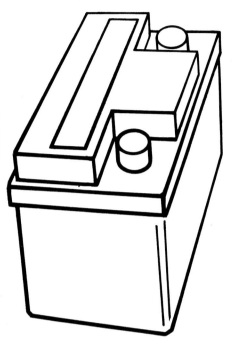

Above: **There are various conventions of line work, although illustrators will also invent specific solutions to the problem of identifying different parts of a structure in line only. These three examples show the same item described in a single line (left); two line weights of which the heavier describes outer edges and parts with open space behind them (centre); variation of line weights along a single line emphasizing the perspective of the drawing (right), the heavy line in the foreground tapering towards the farther edge of the object.**

meshes, wheel-track patterns and similar integral forms.

An alternative option is to use 'perspective line'. Instead of a continuous line of consistent width, the line is tapered to follow the overall perspective of the drawing: that is, the line is heavier in the foreground and tapers when describing receding planes. Sometimes baselines are more heavily weighted to 'ground' the object. These conventions should be used discreetly — too much variation

Right: Cannon
Artist: Michael Gilbert
In this drawing the line widths are judged according to the direction of the light source, giving a cleverly 'shadowed' effect.

Below: Seawater hydraulic 7-function manipulator
Artist: Stuart Michael Molloy
Variation between two line widths (as demonstrated above centre) is subtly employed here to provide clear technical description.

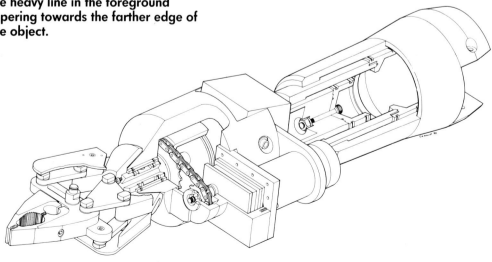

in the visual qualities of the drawing begins to distract from the overall construction and the sense of coherence in the three-dimensional representation. Some elaboration of the image can be applied using techniques for describing tone in black and white, as explained below.

The selection of line widths for inking a particular drawing — that is, the size of the technical pen nibs used

for inking — depends upon the scale of the drawing and the amount of reduction to be applied when it is reproduced in print. If the work is fairly intricate, and the lines too heavy, they are likely to 'fill in' in reproduction — the black lines join up irregularly across the smallest white areas — obliterating the range of detail. If, on the other hand, lines are lightweight in a large-scale drawing,

the strength and impact of the image may be lost.

TONAL VALUES IN LINE WORK

Artwork reproduced as line cannot include solid areas of grey, as these require half-tone processing. There are, however, various techniques for applying tonal values to line work using patterns of black and white which 'read' as grey. This can be done

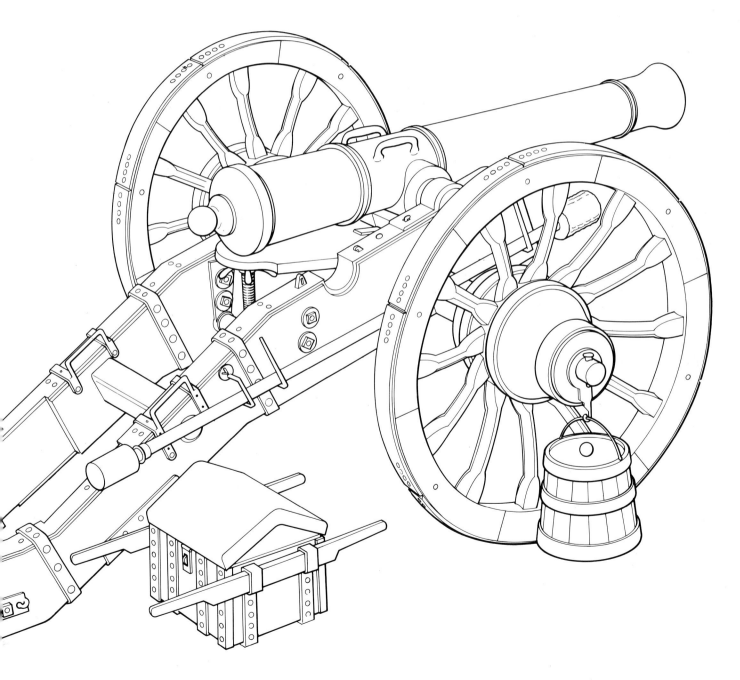

by hand, using variations of line shading and stippling (fine dot patterns), but as contour shading with the pen is time-consuming the alternative is to apply mechanical tints — printed patterns on self-adhesive or dry-transfer sheets. The basic methods of shading line work are described below and the precise effects can be seen in the illustrated examples.

Line shading
Different lengths and thicknesses of line produce a range of tonal effects when arranged to form areas of texture and pattern. The spacing between the lines may be consistent or varied; the lines may be hatched singly or cross-hatched (overlaid in opposing directions); they may follow or cut across the form. Simple line shading which emphasizes contours may extend from or merge into small

blocks of solid black to indicate areas of deep shadow, giving weight and depth to the overall form.

Line shading is either applied freehand, to produce an irregular handdrawn effect, or using rulers and templates to standardize the weight and direction of the lines and provide an even surface texture.

Stippling
The basic technique of stippling consists of applying tiny dots of black over a white area to produce an effect of tonal gradation. Often the stippling is sparingly applied in a drawing which depends upon line quality for its lucidity, giving only a subtle suggestion of volume and contour. At one time, stippling was used to denote specific conventions in technical illustration, such as the edge of a broken section or a surface texture,

but it is now used more broadly for a range of graphic effects.

Small dots widely distributed produce light tone; heavier and more closely spaced dots increase the shadow effect. Because the dot pattern is meticulously applied with the tip of the pen nib, the artist can control the tonal gradation quite finely: some

Right: Inside a nuclear reactor
Artist: Rosie Whitcher
© Science Museum, London
Fine stippling can provide an impressively realistic effect of three-dimensionality.

Below: Oil rig and pipes
Artist: Susan Hunt Yule
Advertising illustration often gives greater scope for interpretive technique. Hand-drawn crosshatching creates a dense tonal effect.

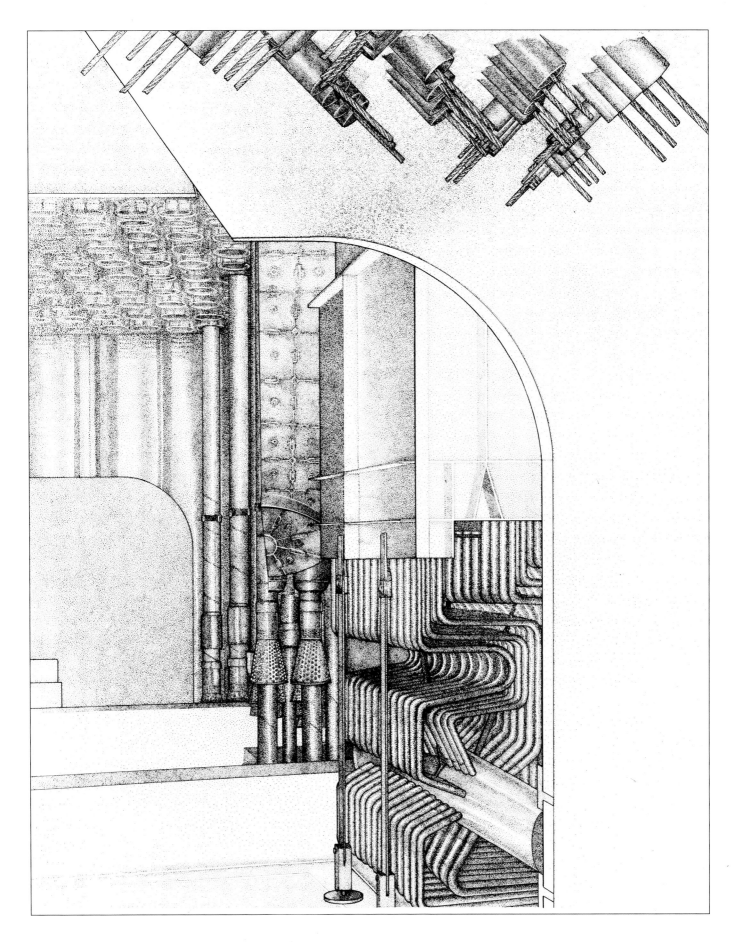

completely stippled drawings have an almost photographic effect.

Beyond the basic conventions of line and stipple technique, the artist may apply his or her own invention to suitable methods of indicating tonal values, extending these to representations of surface texture through not only lines and dots but devices such as small slashes, 'commas' and squiggles which seem to correspond to particular surface effects.

Right-angle drive unit
Ravensbourne College of Design and Communication
A consistent but visually interesting analysis of volume, surface contour and spatial depth is contrived here in black and white only, using ruled line shading, dot and stipple mechanical tints (forming mid-grey tones), and solid black shaded areas.

Mechanical tints
There are three main factors which may lead the illustrator to prefer the application of mechanical tints rather than methods of hand-drawing areas of tone. Firstly, it is a quicker way of covering a large area which needs tonal evaluation. Secondly, the mechanical tint is a non-variable pattern, which may be preferred to the inevitable irregularities of pen-drawn texture. And finally, the overall effect can be seen by looking at the tint sheet, so that its effect on the finished drawing can be judged more precisely before it is applied.

Mechanical tints include a vast range of line and dot patterns, some of which simulate effectively the tonal subtlety of the screened dot patterns formed by half-tone processing of monochrome images. As well as

hatched and radiating lines, and black dots on white, there are graduated tones, white-on-black dot screens and random patterns. The crisp quality and controlled tone of these tints complement the character of precision-drawn line work.

Line and wash
Half-tone and four-colour printing allow the illustrator to employ areas of painted tone and colour in conjunction with line work. Line and wash is a long-established drawing technique which enables the artist to combine the clarity of line drawing with the fluidity of paint media. Washes of tone and colour are employed to provide a more realistic or simply more decorative effect to the final image.

The washes can be applied using either watercolours or drawing inks,

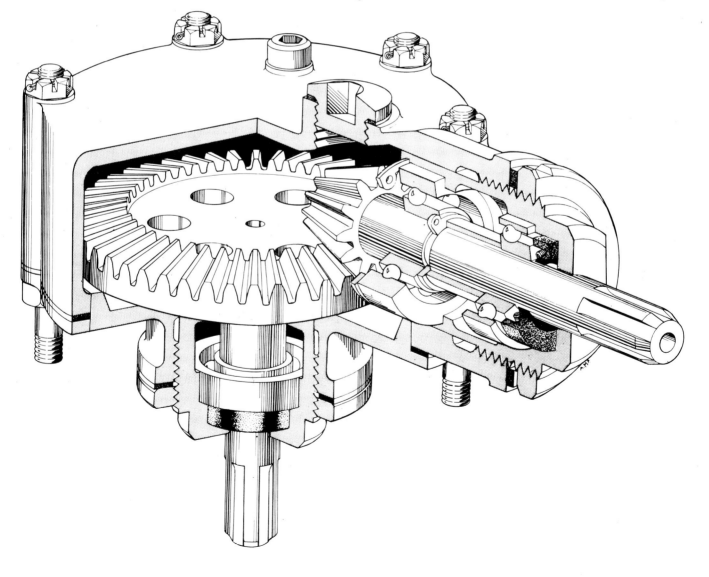

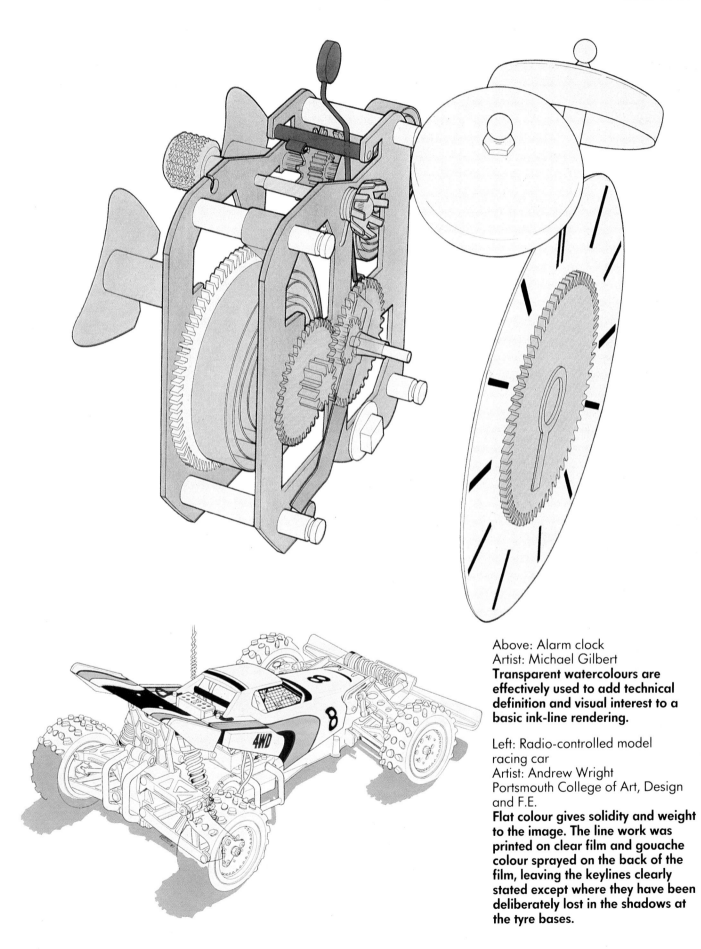

Above: Alarm clock
Artist: Michael Gilbert
Transparent watercolours are effectively used to add technical definition and visual interest to a basic ink-line rendering.

Left: Radio-controlled model racing car
Artist: Andrew Wright
Portsmouth College of Art, Design and F.E.
Flat colour gives solidity and weight to the image. The line work was printed on clear film and gouache colour sprayed on the back of the film, leaving the keylines clearly stated except where they have been deliberately lost in the shadows at the tyre bases.

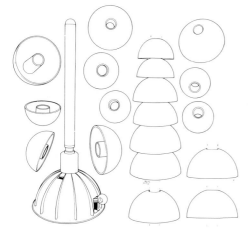

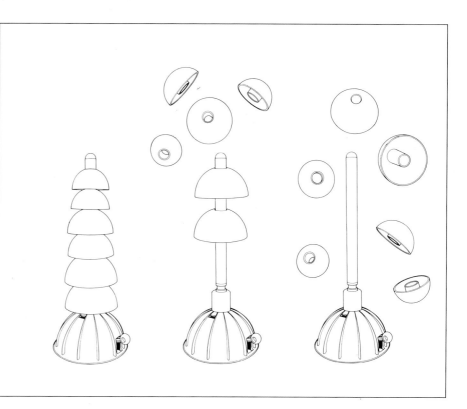

HOLD ALL KEYLINES TO PRINT BLACK.

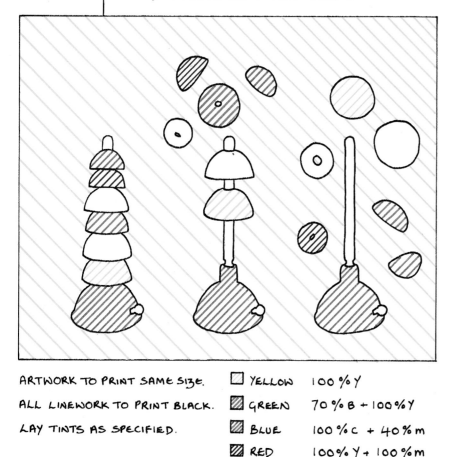

ARTWORK TO PRINT SAME SIZE.

ALL LINEWORK TO PRINT BLACK.

LAY TINTS AS SPECIFIED.

☐ YELLOW	100 % Y
▨ GREEN	70 % B + 100 % Y
▨ BLUE	100 % C + 40 % M
▨ RED	100 % Y + 100 % M
☐ CREAM	10 % Y

Above and left: Line work and colour mark-up for toy package
Artist: Brian Whitehead
Client: David Pocknell's Company for Boots' toys
The child's toy includes variously sized elements of different colours which are threaded onto the stem and 'popped' upwards using the button on the toy base. The design for the packaging, developed from study of the basic shapes in the toy, shows this action in three stages. To print the design in colour, the line artwork is presented with a drawn overlay (left) specifying percentage tints of process colours which the printer will drop into the image as dot screens at film stage.

since these are transparent media and any overlap into the line thickness will not affect its quality. It is an obvious but important point that waterproof ink must be used for the line drawing if wet colour is to be applied over the top. If the work is relatively small-scale, it is easy to produce a smoothly washed effect working with sable paintbrushes. In some styles of drawing, the looseness of the washed tone or colour is part of the particular charm of the image, so that uneven

applications of watercolour which build up fluid tone establish the character of the drawing. If it is important to produce a very flat colour application over a broad area, an airbrush is the most suitable tool. Masking the image (see pages 122–124) allows the illustrator to contain the colour very precisely within the line work.

Overlays
A different method of laying tone or colour into a line illustration involves instructing the printer to introduce tints. These are calculated in percentages of any of the basic colours (known as 'process colours') used in four-colour reproduction — cyan (blue), magenta (red), yellow or black. The artist does not paint in the required colour but provides an overlay of tracing or drafting film on which the tint areas are precisely laid out in relation to the line image. The printer uses this as the guide for processing tint areas on the relevant colour film.

Simple tint patches are usually indicated on a trace overlay by outlining the shape and marking the percentage of colour required — 10 per cent cyan, for example, produces an even tone of pale blue. When preparing mechanicals — full artwork with a line base and overlays — the tints can be applied in various ways. Solid tint patches are either painted in black on drafting film or applied using a self-adhesive coloured film trimmed to shape in place. An irregular shape or

Toy package design
Artist: Brian Whitehead
Client: David Pocknell's Company for Boots' toys
The result of printing up the colour tints is seen here, corresponding to the instructions on the overlay.

variable wash of colour is normally painted on the drafting film overlay. In each case, instructions are provided for the printer on the final colour effect required in terms of process colours.

The designer responsible for briefing the illustrator is usually also responsible for selecting tint colours and marking up artwork before it is despatched to the printer. The briefing process should clearly instruct the illustrator as to the form of presentation required for the work in hand.

DIAGRAMMATIC AND SCHEMATIC ILLUSTRATIONS

Technical illustration is concerned not only with man-made objects, but also with their functions, which may be quite complex. Sometimes the task of the illustration is to show not what an object looks like, but the system or process by which it works or in which it is incorporated. This may vary from a small automobile part which has been developed to function more efficiently than its predecessor, to connected parts of a vehicle performing a specific action, such as the braking or steering mechanisms. Alternatively, an illustration might be required to show the oil flow passing through a vehicle lubrication system, or the stages by which oil is collected from its natural site and piped to a processing plant.

Other examples could be found in any industrial and manufacturing processes or service industries where there is a chain of production involving components, constructions, supply systems and continuous actions. A nuclear power plant, for example, produces energy which is converted for domestic use: illustration helps to explain this process to the lay person. At the other end of the chain, this energy feeds a domestic wiring system which runs all home appliances; these, too, are likely subjects for the technical illustrator.

In such cases, where actions and processes are the subject of the work, the artist needs to devise a diagrammatic or schematic approach which accommodates the scale and sequence of the subject and lays it out in a clearly explanatory form. The illustrator requires, in addition to skills of drawing and rendering in line or colour, a thorough working knowledge of the elements and processes involved, and good design sense which enables him or her to organize the information into a 'readable' visual pattern.

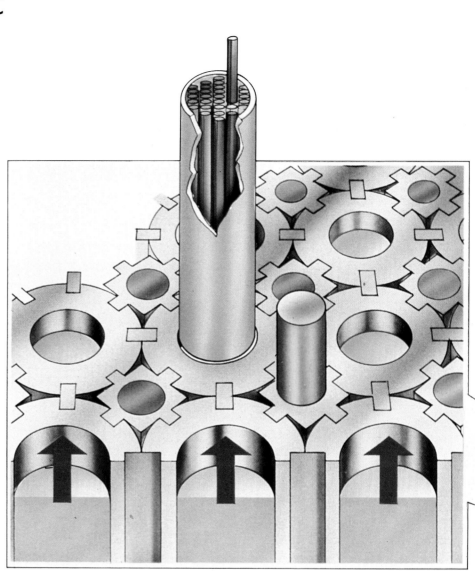

SCHEMATIC INTERPRETATION
A basic factor in this type of work is that the appearance of an object is secondary to its function. This does not mean that the illustrator merely ignores the three-dimensional form and its true dimensions and proportions, but it does mean that the illustration is not bound by them in the same way as in an accurate rendering of a technical object. The relationship of parts, their relative scale and sequence are important. Sometimes these are shown in real terms; at other times it is necessary to approximate them or devise a corresponding scale which indicates but does not imitate the real three-dimensional relationships.

Above and right: Nuclear reactor and core
Studio: Wayne Boughen Design
© UK Atomic Energy Authority
Graphic representation solves the problem of explaining specific elements within a chain of production that emcompasses considerable variations of scale. To demonstrate the construction of the reactor core, it is pulled out from the main image but treated in the same visual style to define its central role within the larger installation. The energy source is shown linked to pylons, drawn in line with an indication of perspective, which suggest long-range continuation of the supply chain.

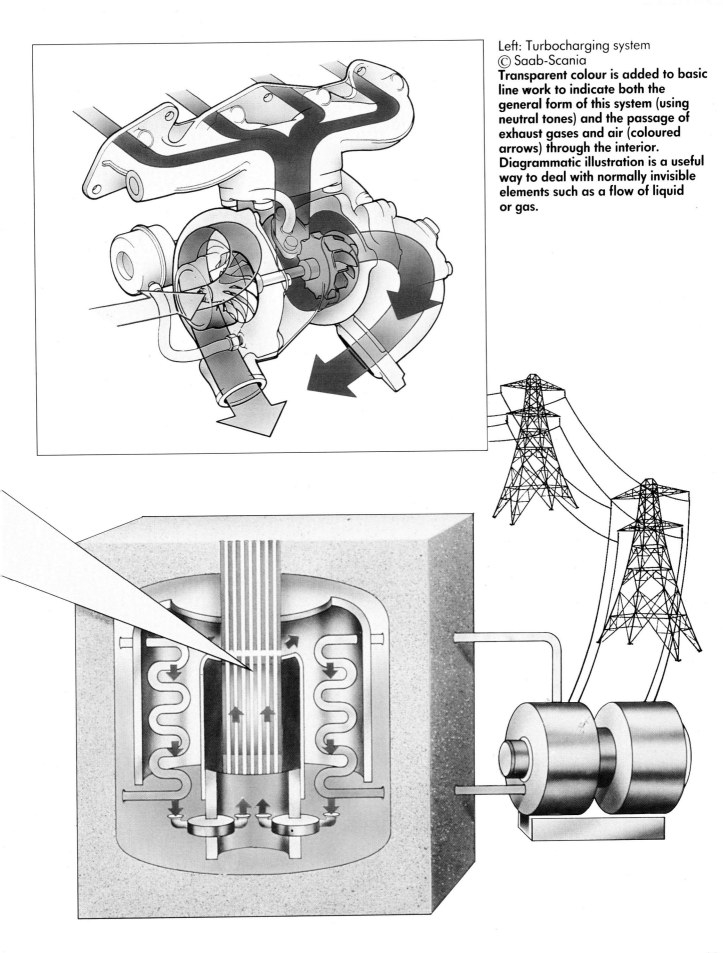

Left: Turbocharging system
© Saab-Scania
Transparent colour is added to basic line work to indicate both the general form of this system (using neutral tones) and the passage of exhaust gases and air (coloured arrows) through the interior. Diagrammatic illustration is a useful way to deal with normally invisible elements such as a flow of liquid or gas.

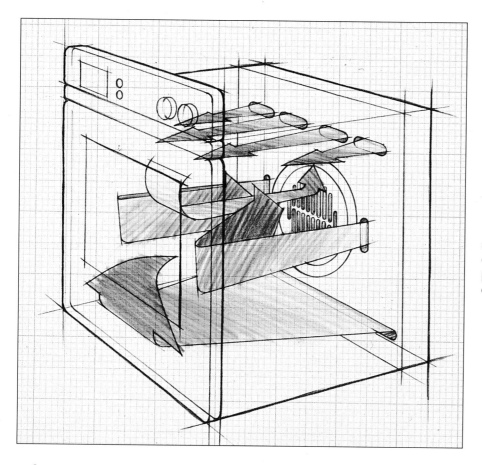

between what is going on at different stages of the process.

A different approach is required in planning an illustration of something which is so complex or on such a large scale that the information needs to be simplified or treated selectively to produce a useful visual interpretation. Wiring diagrams, for example, typically present electrical circuitry as an ordered linear pattern, enabling the viewer to identify and follow individual wires or cables between junctions and terminals. This might be shown as a grid-like pattern with cleanly right-angled bends, for example, and with different supply sources clearly separated. In reality, the wiring

One approach in dealing with a process occurring within a single item, such as the distribution of heat in a microwave oven, is to maintain the general structure in outline or by indication of external planes, in this case basically a box-like shape, and to insert by graphic means the additional information about the internal system or process to be described. An alternative is to 'flatten' the object by taking a simple cross-section to provide the context for the internal activity. A cross-section provides a characteristic shape identifying the object to the viewer without actually describing it; the section sets the framework in which to locate relationships of components and, for example, a fluid or energy source passing through the physical structure.

An advantage of both of these general frameworks is that different aspects of a process can be shown sequentially. The outline or cross-section is used as a common graphic base, appearing each time in the same view and scale. This allows the viewer to make a direct comparison

Above: Multi-function oven
Artist: Mike Charlton
Client: Zanussi Ltd
Arrows defined in graduated colour indicate heat circulation within a confined space. The informal style of coloured pencil on graph paper suggests the type of working drawings which might be produced in the course of a technological design process.

Right: Astra production line
Studio: Grundy & Northedge for Fine White Line Ltd
Client: Vauxhall Motors, Ellesmere Port
Considerable design skill is required in organizing a flow diagram, as well as the ability to make accurate representation of individual illustrative elements.

is probably a spaghetti-like mass with the various paths of the wires looped and crossing, or wires may run at odd angles to accommodate the construction of the mechanisms which they feed. The purpose of the illustration is to present the information systematically, at the same time providing the visual cues which enable the viewer to relate the diagrammatic presentation to the real thing. Similar problems might be confronted in interpreting lubrication, heating or cooling systems, where there are specific conductor channels forming a more or less complex network.

Illustration of a large-scale operation also calls for simplification of the component parts and their connections. This applies to subjects spread over a broad area of physical space or geographical distance: for example, the chain of production in a factory building, or a worldwide communications network. The illustrative conventions will probably become more graphic, further divorced from real form and space, as the range of the operation increases.

The way in which the illustration is laid out also depends upon the scale at which it is drawn or reproduced, and the page format. As an increasing range of information has to be represented within a fixed format, it

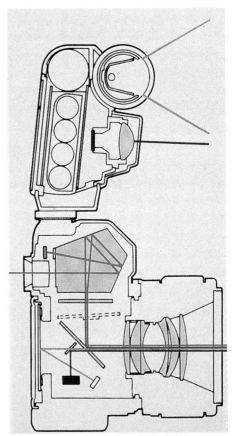

Above: Nikon F-501 diagrammatic cross-section
© Nikon Corporation, Japan
A simple cross-section provides basic information about structure. The uncluttered line work frames the colour elements by which the passage of light through the camera is shown.

may be necessary to abandon even diagrammatic representation of a real item and establish the connections in terms of strictly graphic symbols and typography. Although a technical background may be of advantage to understand and organize the presentation, an illustrator's skills are not necessarily required under such circumstances and this may be treated purely as a design problem.

SOURCES OF REFERENCE
To devise a suitable means of presentation, especially where the essentials have to be selected from a broadly based process, the illustrator needs to know the subject inside out. Unlike a single product view, when a photograph from the required angle might

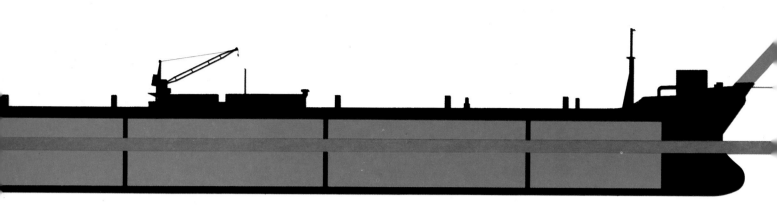

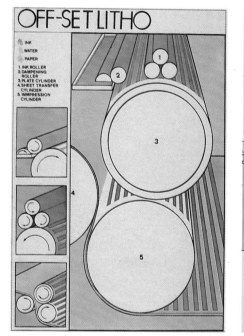

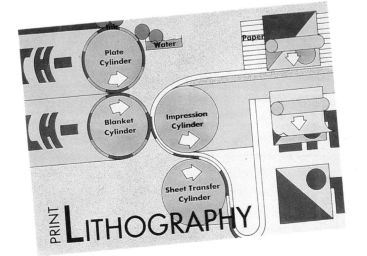

be sufficient reference, for a schematic image the 'invisible' aspects of mechanisms and processes must be fully assimilated in order to devise a graphic equivalent. All information is of value in building a full picture of the subject. The illustrator requires a knowledge of all the many different items that may be involved, in different locations, and the connections between them, such as cables or pipelines, communications or transport systems.

A still wider range of information may also be necessary: maps explaining correct relationships between geographical locations; geological data for industrial processes dealing with natural resources — the illustrator may need to know what is encountered in drilling through the earth's crust, for example, or in conducting a sea-bed exploration. Technological developments normally relate to the physical sciences —

Left: Offset lithography
Final year student project
Portsmouth College of Art, Design and F.E.
These three designs describe the process of offset lithography, in which ink from the printing plate is transferred to the paper via cylinders which produce a right-reading impression of the original image. The same basic elements are represented in each design, but the graphic solutions are differently devised.

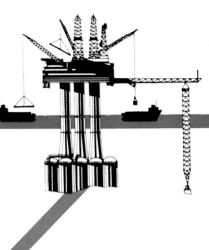

ILLUSTRATION TECHNIQUES

A diagrammatic approach suggests a basically linear interpretation. Much of the work in this area is executed by the same line-work techniques explained in the previous chapter — attention to line thickness in drawing with a technical pen; introduction of tonal values through drawing techniques or mechanical tints; identification of different parts of a system or operation by use of flat or graded colour areas applied by hand-painting, airbrushing or dry-transfer. It may be necessary to include strictly graphic symbols — for example, arrows showing a circulatory action or expressing a direct connection between parts of the image. There are templates and dry-transfer products which supply a range of shapes for such purposes, or they can be hand-drawn in keeping with the style and scale of the illustration.

Broader schematic approaches might make more use of colour, again as a means of identification but also combining effects of tone and colour to establish some sense of three-dimensional depth. The degree to which this is appropriate depends upon the subject and the relative scale of the visual representation. As

Above: Tanker loading
Artists: Michael Robinson/Ron Sandford
Art director: Alan Kitching, Omnific
© Mobil North Sea Ltd
An open design relates the depth of the oil storage point, anchored on the sea bed, to the spread of the operation at surface level.

transmissions of energy in the form of heat, light, electricity, radiation. While it is not generally necessary, nor is there time, to become an expert in matters relating to the subject of an illustration, it may at times be very important to range quite widely in gathering the information to provide full reference for the work.

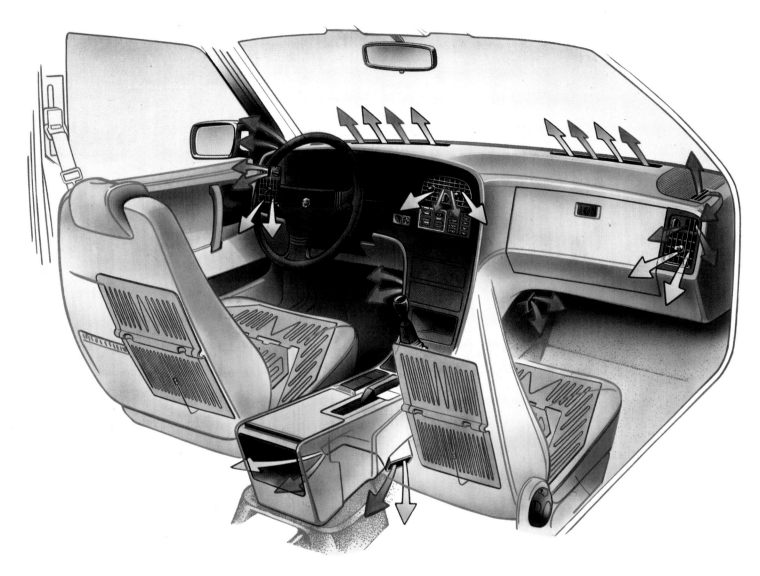

Saab 9000 heating and
ventilation system
© Saab-Scania
**In a semi-realistic rendering of the
car interior, the visual key of
graduated tones is carried through in
colour shading within the graphically
outlined arrows which show where
warm and cool air outlets are sited.**

always in illustration for print, the option to use colour also depends upon the budget for the work, and this must be clearly defined in the illustrator's original briefing. The general approach to interpreting and laying out the information in illustrative form is partly dictated by the range of graphic elements which can be deployed. Line only, or line with black-and-white tonal screens, provides fewer options for visual separation of different elements within the image than does colour work, even where the colour range is relatively limited.

When colour is used extensively in a schematic illustration, it is often applied with an airbrush, which allows the artist to mask off precisely

the areas of the image to be coloured at particular stages of the work. Airbrushed colour can also be laid on very delicately in transparent layers, which can help the definition of, for example, a fluid medium passing through a network of pipes or a pump-action system. Examples of airbrush technique are shown in the illustrations in the following pages; the capabilities and effects of the airbrush are fully explained in section 7.

Specific conventions of technical illustration have been developed to show more about an object than its external appearance, often allowing aspects of its function to be revealed as well as the component parts. Exploded views, cutaways and ghosted images are also schematic representations in the sense that they are designed to open up and explain the construction: these conventions are described further in sections 8–10.

INFORMATION GRAPHICS

The schematic area of technical illustration touches on the borders of the field of design work known as information graphics. This is concerned particularly with presenting hard quantitative information — statistics of various kinds — or sequential relationships, such as a historical chronology or a production system, in graphic form.

The main methods for translating statistical information graphically are standard charting systems — pie charts, graphs, bar charts and tables. These may be constructed flatly or with a three-dimensional element, either in the chart itself or introduced in accompanying illustrations. Typographical content is an important aspect of this work, but there is an increasing demand for lively visual presentation, as statistical studies are often associated with dullness of form and content. This opens up the field to illustrators, who are sometimes required to contribute the more exciting visual elements which the designer may wish to include.

There is much greater scope for illustrative forms of presentation in devising layouts for non-quantitive information. A history of product development in a particular industrial area might, for example, be developed as a charted chronology in which the forms of the product at various stages are fully depicted. Photographic information may be available as illustration, but hand-drawn illustration is often preferred as being more consistent and flexible within the format of the design, following an overall style. It might be impossible to crop photographs to fit the illustration areas of the chart, or the information available in this form might be incomplete.

In a similar way, a sequence of production in anything from soft-drinks manufacture to vehicle assembly-lines can be presented as an illustrated flow diagram. The sequence of the design follows the process of manufacture and hand-drawn illustration allows each stage of the product to be shown graphically, in any mode from a simple line diagram to a colour rendering on limited scale. This sort of illustration is not the sole province of technical illustrators, but by the nature of the subjects a technical background is again often of advantage.

Cash dispenser
Artist: Michael Gilbert
Design: Wolff Olins
Client: Midland Bank plc
Simple line diagrams are a clear and economical way of demonstrating a specific process, as in instructing the customer in the use of an automatic cash dispenser for bank deposits and withdrawals. The hand sets a uniform scale within each illustration which helps to clarify the step-by-step procedure.

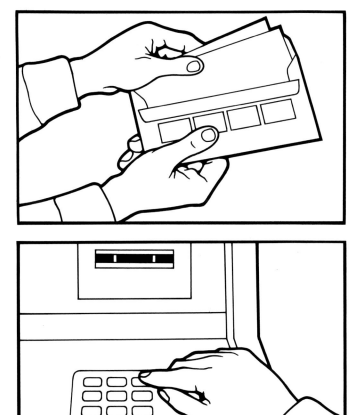

COLOUR RENDERING

The process of preparing a detailed drawing as the basis of a colour rendering will have provided information about tonal and colour values which the artist can put to use in developing the final image. As with drawing, an effective representation is achieved by analysing the visual impression of the object very closely and finding the best means to translate this analysis into a colour medium.

TONAL AND COLOUR VALUES

Colour is a physical attribute of a material. Tone is a relative scale of values representing degrees of illumination, which can be demonstrated graphically as a progression from black through a range of greys to white. This basic tonal scale is colour neutral, but is used to interpret patterns of light and shade falling on a surface or object which has inherent colour values; the pattern of illumination describes surface qualities and three-dimensional form. There is a continuous interaction of tone and colour which describes all observable phenomena in terms of colour, pattern, surface texture, contour, volume and spatial depth. These are the elements of representation which have to be identified and applied to a colour rendering.

ILLUMINATION OF VOLUME

Tonal relationships provide depth cues. Taking the example of a simple cube, an outline perspective drawing

Car engine
Artist: Bob Freeman
Art director: Adam Stinson
Client: F & F Publishing
A vigorous, painterly style of watercolour over pencil drawing produces a technically detailed image with the spontaneity of a sketchbook reference, but tonal and colour values are precisely rendered in the energetic brushwork.

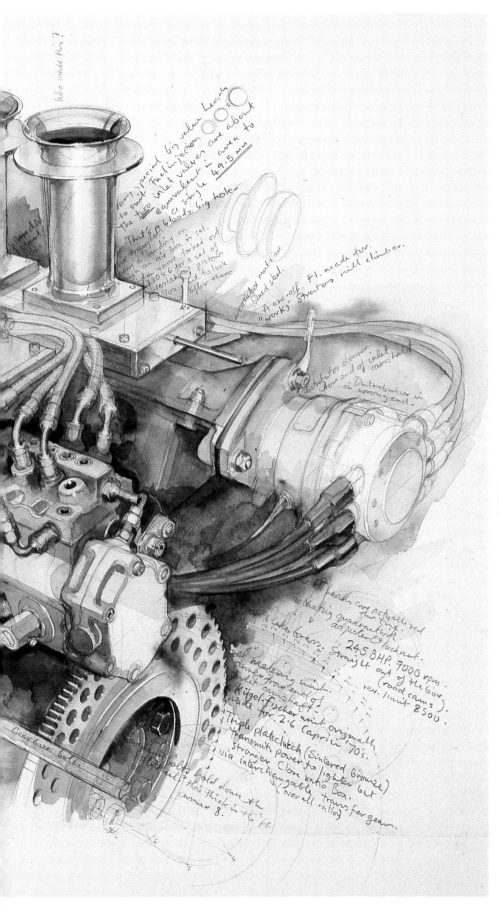

Consistent effects of light and shade lend form and character to colour renderings. In these examples the light is falling from upper left making similar distribution of shadow and highlight but in different degrees: even light with a balance of tonal values (top), an effect of strong lighting with high tonal contrast (centre) and soft light diffusing the contrast (bottom).

describes the planes of the solid form. The sense of solidity and depth is enhanced by laying in, for example, a dark tone on the right-hand plane, a mid-tone on the left, and a pale tone across the horizontal top plane. This corresponds to illumination falling on the cube from a source above and to the left of the object, leaving the right-hand plane in shadow because it receives no direct illumination. The same general pattern of tonal change from left to right and across the top of the object will occur in any object, whether it is a regular geometric solid or a more complex shape, if it is placed under the same light source.

The establishment of a real or presumed light source is therefore an essential basis for the illustration of a technical subject, since the pattern of light and shade models the three-dimensional form. Either the object is positioned as required in relation to a directed light and the effect is analysed by observation, or a light source is assumed as a fixed point and every aspect of the rendering

consistently related to the spread of light from that point. The example of the cube demonstrates a convention of technical illustration in which the light is presumed to fall on the object at an angle of 45° or more, from a position above the observer's (or artist's) left shoulder. However, the illustrator is at liberty to fix the light source in any position which enhances the rendering of a given three-dimensional form.

The pattern of light and shade caused by a specified light source must be followed consistently throughout the rendering to clarify the forms. Variations in the pattern destroy the coherence of the image. While this might be acceptable in an editorial illustration intended to induce an atmospheric effect, it is not appropriate to a technical rendering aiming, for example, to locate and describe machine parts. A network of cylindrical components will be quite unreadable if every cylinder has a random distribution of highlight and shadow. Dramatic tonal contrasts

also tend to suggest a mood rather than straightforwardly describe a form. This is generally undesirable in technical illustration, which usually aims to be strictly representational, but occasionally in sales and advertising contexts, the visual impact of the image can be appropriately played up by manipulation of the tonal contrasts.

A basic sense of three dimensions is gained through application of simple tonal blocks; smooth tonal gradations supply a more sophisticated portrayal of real space and form. This is especially the case in technical illustration, where the qualities of both the component materials and the product design are highly developed and this qualitative standard is often visually apparent in a notable sleekness of form and surface finish. In illustration work, polished surface effects are particularly associated with airbrush painting (see the following section) but are also achieved by a variety of different techniques.

The tonal pattern in a complex

Correct plotting of cast shadows, with a fixed light source related to the original vanishing points, adds to the authenticity of a rendering. The different spreads of the shadow image are shown with a conventionally positioned top-left light source (below), low-angled light casting a long shadow (below left), and direct overhead light (left).

Ford Sierra
© Ford Motor Company Ltd
Even quite minimal colour work can enliven an image and give it weight and solidity. Markers are a useful medium for rapid colour elaboration. A combination of warm and cool greys enhanced by touches of blue and red makes an effective contribution here.

image is contrived, within the framework of the consistent light source, to clarify important details of three-dimensional form by providing appreciable contrasts between changing planes, degrees of curvature, and solid or hollow forms. There is a continual movement from light to dark, through degrees of mid-tone, which subtly emphasizes edge qualities and breaks up the painted surface so that the viewer is made aware of multiple components and the spaces they occupy. In a broad exterior view of an object, however, there may be little spatial recession and the tonal changes may be minimal and widely spread. A distinct but highly controlled emphasis on the lightest and darkest elements of the image can sharpen the rendering without over-exaggerating the form.

CAST SHADOW
The tonal qualities so far discussed are related to shadows describing volume. Cast shadows, on the other hand, give objects a specific location in relation to surrounding planes and adjacent objects. The simplest example of cast shadow is a silhouette of the object thrown onto a vertical or horizontal plane. Especially in a rendering of a single object standing on a horizontal plane, a cast shadow emphasizes the ground plane and prevents the object from appearing to float. Graphic conventions for show-

ing cast shadows range from solid, hard-edged blocks of dark tone to vignetted graded tones.

To plot the extent and shape of the cast shadow in relation to the object, the artist must establish a position for the light source in relation to the picture plane. This becomes a fixed reference point: straight lines drawn from the light source to pass through points on the outline of the object continue on to the ground plane and define the corresponding points at the outline of the shadow. A low light source forms an elongated, faint shadow; a high angle of harsh illumination creates a truncated solid-toned shadow. This is analogous to the effects of sunlight at late afternoon or midday respectively, although in much technical rendering, light sources will be artificial or even imaginary.

Cast shadows are also thrown by one object on to another. In a complex assembly, such as an automobile or aeroplane engine, the many components within a given space are forming subsidiary patterns of cast shadow. Sometimes one part throws a shadow right over a component lying behind it which, if rendered faithfully, would appear to distort the shape of the second item. The artist may prefer to adjust the shadow in the interests of producing a clearly defined image. On the other hand, the shadow 'outlines' formed by, for example, the overlapping blades of a rotor con-

struction can make the image appear more crisply three-dimensional.

REFLECTED LIGHT
Many of the materials used in manufacture of technical objects have a reflective quality — metals of all kinds, plastics both opaquely coloured and clear, and glossy paint finishes. Reflected light is another element which provides focused clarity in an image. Line highlights representing reflected light can, for example, be applied to the leading edges of receding planes. Fine details such as a screw head or bolt show more precisely against darker-toned areas if the edges are highlighted with lines, dots or dashes of white.

In curving forms with the curve turning away from the viewer, the darkest-toned shadow rarely extends right to the edge farthest from the light source, as there is likely to be some light reflecting back from the plane behind which slightly illuminates the outer contour. In a similar way, in a circular or elliptical form with a hole cut through the centre, the lower curve of the inside ring may have a highlight at the far edge where light is thrown back upon it from the other side. Careful reproduction of such effects of reflected light give authenticity to a rendering.

These are quite subtle examples of uses of reflected light effects. In general terms, the more reflective a

Above: **The surface qualities of different materials are shown by variations in tonal contrast and colour values. A matt finish has subtle tonal changes (left), a shiny surface displays hard-edged tonal contrasts (top right). Reflected colour is often conventionally shown as cool blues set above warm earth colours, suggesting sky and ground.**

surface is, the more pronounced are the tonal contrasts. This is a principle constantly in use in technical illustration to define different qualities of surface finish.

LOCAL COLOUR

This term refers to the actual colour of a surface, object or material. To visualize this, imagine a smooth plane surface evenly lit by white light: the colour it shows under these circumstances is the local colour. Under normal conditions, local colour is an important aspect of a rendering of a three-dimensional object, but it is modified by tonal influences which describe the form. These can be represented by a range of tones of the specific colour: for example, a solid, bright red passes into light red where it is reflecting the most light, and into a

dark red in the shaded end of the tonal scale.

This is, however, a relatively simplistic interpretation which merely illustrates a basic principle. There are many other influences which may affect the way the colour appears on different areas of an object. The subtle changes of a single colour affected by light and shade and the interactions between different colours need to be identified and used to enliven a rendering, unless the treatment of the subject is deliberately flat or low-key. Even quite mundane or visually unexciting objects may reveal some surface effects which translate into interesting detail in illustrative terms.

REFLECTED COLOUR

Colour is an effect of light, and a reflective surface sometimes shows unexpected colour elements together with the tonal range representing light and shade. The more shiny the surface, the more clearly the reflected detail appears, so that chrome, for example, typically displays a more

Below: Renault
© Document Renault
Reflected colours are used to 'locate' an image, as here where blue tones, as from the sky, are reflected in the top of the vehicle and warmer earth colours in the lower bodywork.

complex effect of surface tone and colour than steel or aluminium. A familiar graphic device is the complete mirror effect, in which immediate surroundings are fully reflected perhaps in distorted shape, as in the hub cap of a car wheel. This is an extreme case, and one which makes for visual interest; but typically reflections are much more subtle and more abstract than a definable mirror image.

A basic example of use of colour reflection might occur in the rendering of a polished metal pipe running alongside a painted metal casing: the colour of the paintwork will appear somewhere in the reflected detail on the shiny metal. This adds not only an element of realism to the rendering, but also forms a visual link between the components which makes the illustration more coherent to the viewer. Incidental colour elements of this type are useful illustrative material.

Specifically related to technical illustration is a convention of creating an effect of reflected colour by locating an imaginary horizon line on an object, above which the colours tend to blue, as if reflecting the sky, and below which they are tinted with warm brown or yellow earth colours, as if

Left and right: Ballbearing and toothed wheel
Artist: Tom Steyer, The Garden Studio
Reflective materials offer many hints of colour values. Emphasis and enhancement of colour differences can add to the image quality without detracting from technical analysis of an object's construction.

reflecting the ground. This principle is often applied to automobile illustration, including cars, trucks and motorbikes. The sky/ground colour division is particularly used in rendering highly reflective chrome parts, combined with the appropriate effects of tonal contrast. It is a useful device for working out details of reflected tone and colour where the object is not given a specific location or surroundings that would otherwise supply these details.

The horizon-line idea is also interpreted tonally, as in rendering the glossy paintwork of a car body. The line might be established, say, at about the level of the wheelguard mouldings: above the line tones are generally lighter, leading to areas of strong highlight, and below shadows are heavily emphasized, on the principle that the sky throws illumination on the object while the ground area reflects back little light.

This convention is not only applied to sizeable objects, such as large vehicles, which are usually seen in the open. It can be applied quite subtly to smaller objects, including machine parts, engines or domestic appliances, to enhance the visual detail of the surface effects.

INTERACTION OF COLOURS
The sky/ground colour combination also demonstrates two basic principles of colour theory. Blue is a cool colour whereas earth colours are warm. The opposition of warm and cool colours sets up a sense of contrast and emphasis in a painted image. This can be extended across the colour spectrum: blues and greens generally tend to be cool colours, while reds, oranges and yellows are warm; a blue-purple is cooler than a red-purple, and so on. Cool colours give an impression of space and distance, while warm colours tend to project forward. These attributes, while they need not be used in an exaggerated way, help to establish the perspective and spatial depth of a colour rendering. For example, a strong red at the farthest edge of an object will seem to come forward, conflicting with the drawn perspective. Red needs to be given a more muted

Prototype cars
Artist: (top) Julian Thomson
Studio: Lotus Design
Artist: (above) Mathias Kulla
Tonal contrasts affect the mood of an image, as does the choice of strong or neutral hues in the colour work. A coolly sleek image is conveyed by selecting muted blue and brown tones together with black and neutral greys (above), while heavy blacks slashed with red and purple give dramatic impact and a sense of motion (top).

tone, or even a slight bluish tinge, the farther from the viewer it appears.

The second element of colour theory is the mutual enhancement of opposite colours and harmony of similar colours, or those which have common elements. Yellow and blue are mixed to make green, therefore green merges naturally with yellow or blue but is the opposite, or complementary, colour to red because it contains no red. Other complementary pairs are yellow and purple, and blue and orange. Directly opposite colours placed side by side tend to make each other more vivid, whereas

related colours form a harmonious gradation. These are interactions which again can be discreetly employed in an illustration to give vitality to an image.

In technical illustration, neutral shades are frequently employed to render metals and even paintwork. By the nature of the objects which the illustrator needs to depict, combinations of bright colours are untypical. Black, white and a vastly subtle range of greys commonly occur, together with beiges and creamy tints, in metalwork, plastics and paint finishes. The introduction of very controlled colour

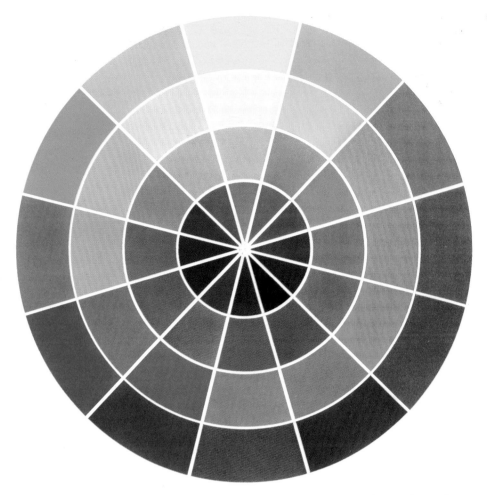

Above: **A range of coloured greys and neutrals can be mixed from black and white with colour tints (top line and left-hand side), from pure colours (diagonally from upper left) or by further combining these mixtures (left and right of diagonal).**

Left: **The colour wheel shows the relationships of primary and secondary hues; for example, red merges into orange on one side and purple on the other, but is opposite the range of greens. This rendering shows true colours in the outer wheel and tonal values in the inner rings.**

suggestions helps to enliven the surfaces: metals, for example, rendered basically in monochrome, can be carefully studied to see whether they convey cool or warm tones; a little blue or brown mixed into a grey will give it greater visual depth and interest. But colour cues must be used with restraint, otherwise the combined effect in a rendering is unrealistic.

RENDERING TECHNIQUES

With the opportunity to use full colour in a technical illustration, the artist is confronted by a range of decisions about the style and execution of the work. Of particular importance to professional illustrators is the time factor, which might tend to favour the rapidity of marker rendering over the meticulous technique of coloured pencil drawing, for example, and gives water-based media such as watercolour, gouache and acrylics a clear advantage over slow-drying oil paints. The characteristic surface qualities of the medium also play their part: there is no reason to attempt to work in pastel if the image requires pinpoint accuracy; it is less easy to produce broad, flat areas of bright colour with the transparent medium of watercolour than with the substantial opacity of gouache.

In the commercial world, an increasing number of freelance illustrators are competing for a more or less limited amount of work. It is common for an individual artist to develop a particular style which identifies his or her body of work and provides prospective clients with a point of reference when considering suitable illustrators for particular projects. This style often relates closely to the medium employed and in practice many illustrators depend upon a relatively small range of techniques. The demands of technical illustration, too, tend to restrict the range of media and methods. There is not much room for the kind of quirky inventiveness which can be applied in the broader field of advertising imagery, when the style is intended to become a selling point in the campaign. Technical illustration often requires extreme precision and a technique of almost photographic realism.

The following is a summary of basic techniques and the characteristic effects of specific media for colour rendering. It is intended both as a guideline for practical application and a signpost to personal experiment with the available illustration media. Although technical illustration is not a field that encourages visual risk-taking, there may be occasions when a fresh approach is needed and the artist can investigate the different possibilities of an unfamiliar medium. For students of technical illustration, there is also time to make these investigations before the pressures of

The subtle transparency of watercolour is employed in a restricted range of warm greys to render the metallic casing. In keeping with the texture of the medium, the numerals on the keys are defined with softly brushed edges.

professional practice begin to exclude a free-ranging approach. There is no single correct technique for using any medium: a method which produces the required effect is the right one for the job in hand.

WATERCOLOUR

This is one of the most commonly used of the illustration media. It is versatile in effect, ranging from soft graduated washes of colour to tiny, hatched, jewel-like coloured marks. The main quality of watercolour is its transparency: it is a fluid medium and spreads easily, but if there is any distinct mark on the surface underneath the colour application, this will show through to a greater or lesser extent. The colours gain luminosity, however, because of the white underlayer provided by the surface of the paper or board.

Techniques

There are various different techniques for applying watercolour. Flat colour is built up by working evenly back and forth across the surface. Because the paint is water-soluble, the brush-strokes can be blended together; they need not form hard edges or streaks if the medium is properly handled. Suf-

ficient colour to cover the whole area should be mixed to the required dilution before starting, and the brush should be recharged frequently to keep the paint layer damp and even. Watercolour lightens as it dries, so the wet colour layers need to be built more strongly than is required in the final effect.

Soft washes, in which the colour graduates from the pure hue to a pale tint, are created by working wet into

wet, building up the darker tones with overlaid colour, or wet over dry, which creates more intense colour or tone. To spread a wash evenly on a rough-surfaced paper or board, the surface may be wetted out with clean water before the colour is applied.

Watercolour is a suitable medium for applying flat or graduated colour washes into line work, because its transparency leaves an ink line unaffected if the colour overlaps the thickness of the line, whereas opaque gouache colour covers and breaks the line. The line ink must, of course, be waterproof or it will spread into the colour. Watercolour washes also work well when applied to pencil drawings, having a delicacy suited to the subtlety of graphite blacks and greys. When the colour is dry, the drawing can be reworked in pencil, creating a layering process of building the detail in the image.

Watercolour washes can either be controlled within a clearly defined delineation or can be flooded into a

LAYING A WATERCOLOUR WASH

1 To obtain an even wash, it may help to wet the paper first. Use a broad brush to lay colour smoothly.

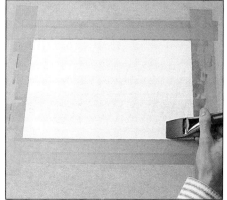

2 Work from side to side across the paper, travelling downwards to complete the overall wash.

LAYING A GRADUATED WASH

1 As before, start at the top of the paper and lay the colour in broad bands working from side to side.

2 Continue to work downwards, painting wet colour into wet so that the brushstrokes continually merge.

3 To fade out the colour, further dilute the paint with water and brush over the edge of the previous stroke.

drawing to create a loosely worked, sketchy effect. A completely different method, comparable to the traditional technique of painting in egg tempera, involves applying the colour in small brushstrokes of more solid hues. This creates a luminous colour effect: small dashes or dots of colour, or linear hatching can be woven into vibrant colour mixtures which blend in the eye but provide a richly textured surface. The paint should not be much

diluted, as in successive applications the excess wetness will dissolve and lift previously applied colours.

In a relatively small-scale and delicate rendering, watercolour wash techniques may be inappropriate to the scale of the colour areas. A detailed painting can be built up by applying the colour in a whole range of different ways, using whatever technique is suitable to produce the required effect of form or texture. The

Above: 1920s steam-powered portable threshing machine
Artist: Graeme Chambers
Washes are the basis of this watercolour rendering, gradually increasing in strength to build the colour density. Textural detail was worked over the broad colour areas. No white paint was used: the surface of the white CS10 paper stands for the highlight areas and adds brilliance to the pale tones.

Left: Turbo car engine
Artist: Bob Freeman
Art director: Gordon McKee
Client: Creative Business for Citroën
UK Ltd
**Details of form and surface texture
are obtained by building up
watercolour washes in patches and
overlapping layers to produce
complex effects of tone and colour,
using the brushmarks descriptively.**

paint effect will still be transparent,
however, and this should be taken
into account in considering the
sequence of the work, especially
whether pale or dark colours are laid
in first.

Stippling of colour is usually
applied by dotting the surface using
the tip of a fine paintbrush. Alternatively, for a rough stipple, wet paint
can be loaded onto the bristles of a
toothbrush and flicked with the
fingertip to create a shower of coloured dots. It may be necessary to
mask out the areas surrounding the
stipple texture, using paper masks
and low-tack tape or masking film.
Previously painted areas should be
allowed to dry completely before
masking is applied, otherwise the surface effect will be damaged.

Drybrush technique produces a
roughly linear effect of broken colour.

▶ *WATERCOLOUR TECHNIQUES*

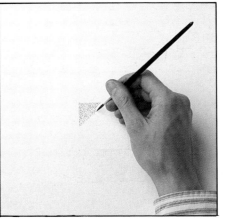

Stippling **Use a fine brush to apply
dots of colour. Tones are varied by
spacing the dots widely or closely.**

Drybrush **Load the brush with colour
and blot it, then spread the bristles to
create a linear, dragged texture.**

Spattering **Apply wet colour to an old
toothbrush and flick it from the
bristles: masking may be needed.**

Hatching **Draw straight lines of
colour using a fine brush guided on
the edge of a ruler.**

Left: Prototype car
Artist: Peter Byatt, Spectron Artists
Client: Peat Marwick McClintock
In a descriptive image created for the client's corporate brochure, the fluid medium of drawing ink is applied to make effective contrast between rough natural textures and the smoothly highlighted vehicle.

Right: Industrial complex
Artist: Peter Byatt, Spectron Artists
Client: Peat Marwick McClintock
Washes of ink are both crisply edged and softly blended to provide a vivid contrast of light and shade across this industrial view. The effect is enhanced by consistent offsetting of warm yellow against cool blue tints.

effect. Water-soluble inks are more directly comparable to watercolour but the textural range is slightly restricted, as ink is always liquid, whereas watercolours also come in a more viscous consistency.

Drawing inks are employed to draw coloured lines, using a dip pen or ruling pen (not a technical pen, which requires a specially formulated ink). Linear work can be used to crispen washed areas and enhance tonal contrasts.

The brush is loaded with paint, blotted with a tissue and the bristles spread to apply the residue of colour in fine scratches and trails. These textural effects are useful for simulating natural materials such as wood or stone. On a smooth surface, it is also possible to scratch back highlight detail, as explained in the section on airbrushing (see page 126).

DRAWING INKS

Coloured inks can be used in broadly similar ways to watercolour, for applying flat, transparent colour blocks, loose washes or effects of surface texture. The colour range in waterproof inks is relatively limited but the colours appear generally more intense, though because they are immovable when dry there is less scope for reworking the surface

USING A RULING PEN

1 Charge a paintbrush with ink (or diluted paint) and insert colour between the two blades of the pen.

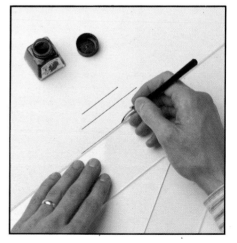

2 Hold the pen at a slight angle and draw the flat blade smoothly along a straight edge to rule the line.

GOUACHE

An opaque medium offering a brilliant colour range, gouache is the main alternative to watercolour as a favoured medium for technical illustration work. The opacity in the paint comes from a white filler substance, so washes of gouache do not have the clarity of watercolour washes. If built up layer over layer, the colours quickly become muddy. The primary effects of gouache are areas of solid, flat colour, even gradations of tone and heavy shadow effects offset by opaque white highlights.

Techniques

The filler in the paint encourages gouache colour to dry flat, so it is easy to achieve an even spread of solid colour. Tonal gradations, unlike with watercolour, are made by blending mixed shades and tints together while the colour is still wet. Very fine details such as lettering, linear detailing and sharp highlights can be crisply painted with a fine sable brush. If not overdiluted, an opaque white or pale tint will sit cleanly even on a black ground for effects of highlighting. The tonal range of a gouache painting can be subtly controlled or intensely contrasted.

Dried gouache is not waterproof, but it does not lift easily. Because light colours can be worked over dark or vice versa, continual adjustment can be made to colour and tonal values by overpainting. Some of the stronger

Stippling **The opacity of gouache means that stippled colours make a distinctly 'spotted' effect.**

Spattering **Using a toothbrush to spread uneven dots of colour creates a rough stone-like texture.**

Drybrush **The broken colour effect of dragging the paint forms an uneven linear texture like woodgrain.**

Hatching **Broad bands of colour are used here with different widths and spacing to vary the tonal values.**

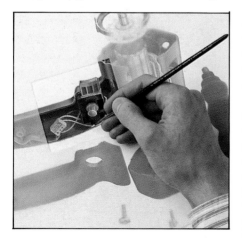

Highlighting **Because of the opacity of gouache, pure white highlights can be painted over dark colour.**

colours have an occasional tendency to bleed through an upper layer, but if an area is allowed to dry thoroughly before it is reworked, this should not be a major problem.

The ability to overwork gouache makes it an ideal medium for developing textural effects by careful simulation of minute changes of colour and tone. The broken colour effect formed by the drybrush technique, as explained in the section on watercolours (see pages 103 and 104), produces a more solid and substantial effect through the opacity of gouache, and complex qualities of mixed tone and colour can be achieved. Stippling and hatching are also suitable techniques for work in gouache.

Gouache is often applied quite thickly. Solid lines of colour can be ruled by guiding a fine brush or ruling pen along the edge of a ruler (raised to prevent colour bleed). Even lines may also be drawn freehand using a fine square-tipped brush, if it is manipulated with a steady hand. Clean-edge qualities are also a characteristic of gouache, as it can be worked evenly around an outline and blocked in across the remainder of the shape. Care should be taken not to overload the brush unless the textured effect of the brushstroke is a required element of the finished surface: a thick ridge of colour forms

BRUSH RULING WITH GOUACHE

 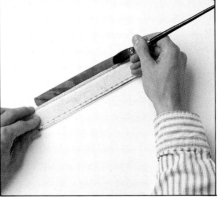

For a fine line, use the point of a small round-tipped sable brush and guide the brush along a ruler edge.

For a solid band of colour, work in the same way using the ruler to guide a broad flat-tipped brush.

Below: Cable-laying ship
Artist: Nick Trudgian
Studio: Blue Chip Illustration
Client: Cable and Wireless
Brilliant colour is characteristic of the opaque medium of gouache. The ship is described with meticulous detail, the pattern of light and shade on its white-painted sides and rails emphasized by the use of cool blue in the shadows. This contrast is repeated in the foaming waters in the wake of the boat and enlivened by green tints, again visually linked to colours in the boat deck and paintwork, giving additional depth to the heavy sea blues.

Blast furnaces, Kwangyang
Steelworks, Korea
Artist: Philip Crowe
Engineered and constructed by
Davy McKee
**Constructed from engineering
drawings supplied by the client, this
industrial image offers an
opportunity for atmospheric effects
as well as technical detail. The
opaque gouache is applied both as
solid colour and as loose washes.**

along the edge of the stroke when the
brush is heavily charged.

Gouache dries to a matt finish
which is easily marked by dirt settling
on the surface and fingerprints or
other greasy marks. A sheet of clean
layout paper should be used to protect previously worked areas of the
illustration while close detail is being
applied.

Gouache and watercolour are also
the main media used in airbrush
rendering. Details of airbrushing technique are given in the next section.

ACRYLICS

As the most recently developed of the
available paint media, acrylics were
for some time used in ways designed
to imitate the qualities of oil paint,
being both more economical than oils
and less laborious to use. The range
of acrylic paints has, however, been
further developed and investigated as
a distinctive medium in its own right,
and is nowadays more effectively
exploited in illustration work than was
previously the case.

Techniques

Techniques used in watercolour and
gouache painting can be applied to
acrylic paints, but the results are
somewhat different. Thinned with
water, acrylics provide transparent
washes of colour which dry to a waterproof finish. Thin layers of colour can
be built up to considerable density or
given crisp hard-edged qualities
which distinguish them from the softer
effect of watercolour washes.

Acrylics may also be treated in a
manner borrowed from traditional
methods of oil painting, laying in dark
shadows and mid-tones as an underpainting and colouring up by the
application of thin glazes. The plasticized composition of acrylics tends to
form a slightly glossy surface finish,
and matt medium is sometimes mixed
with the paint to reduce this quality.
The slight sheen is useful when the
glazing technique is used, however,
as layering of colours diluted with
water creates a deep, rich, glazed
effect. Alternatively, used straight
from the container, diluted only
enough to allow them to spread
easily, acrylic paints can provide
areas of solid colour (with none of the
slight chalkiness typical of gouache),
broken texture and even thick
impasto.

However they are used, acrylics,
are quick-drying and once dry cannot
be lifted effectively; revisions are
made by overworking. The paint dries

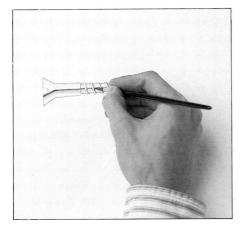

Underpainting **As acrylics dry to a waterproof finish, colour glazes can be laid cleanly over shadow detail.**

by a bonding action due to its polymer base, not by simple evaporation of the liquid content, so the colours remain truer as they dry than is the case with watercolour or gouache.

Acrylics tend to be more translucent than gouache but not as transparent as watercolour, though the degree of transparency or opacity varies and may be related to the intrinsic quality of the colouring matter in the paint. It may take two or more applications to achieve an area of solid, flat colour. Mixing white into the paint is a good method of increasing opacity, as long as this does not let down the colour value too much.

Textured qualities can be achieved by weaving patterns of stipple, hatching or flicks and dashes of paint, as with watercolour, and by using drybrush technique. Acrylic paint can be built up very thickly, as the dried paint layer is relatively flexible, and impasto effects are readily achieved if a heavy texture is required. The paint

Below: Refuelling
Artist: Mike Vaughan
© Mobil Oil Company Ltd
The heavy consistency of artists' acrylics lends itself to bold brushwork and strong colour values. This illustration has a clear narrative element, but also a powerfully abstract design sense.

Bicycle saddle
Artist: Tom Steyer, The Garden
Studio
Oil paint is not always practical for illustration work, as time needed for preparation and allowing the paint to dry can be prohibitive. The medium contributes, however, a particularly rich luminosity and density of colour which enhances the illustrative quality of an image. Fine glazes, softly blended hues and thickly opaque dabs of colour merge into lustrous surface effects.

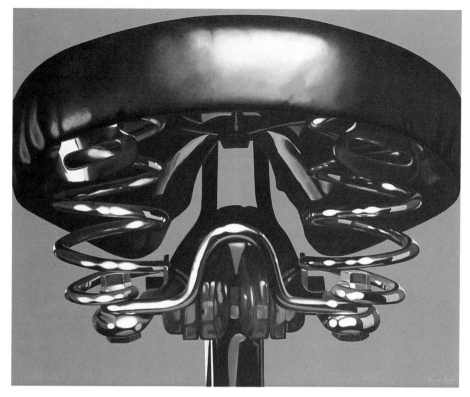

can be laid in thickly using a brush or a palette knife. It remains workable longer than gouache, but once it has begun to dry it cannot be scraped back easily, as it forms a rubbery skin of colour which leaves a ragged edge if peeled away. If the paint has been very thinly applied it may be possible to scratch back small highlight areas with a scalpel (utility knife) blade, but this is less necessary than with water-colour or ink. Instead, a white highlight can be painted in using acrylic; as this does not have the slight chalkiness of gouache, it does not risk spoiling the coherence of a semi-transparent colour effect.

OIL PAINT
As has been explained, oils are rarely used for illustration work because the additional time needed both for preparation and to allow the finished work to dry out completely makes it commercially impracticable. The main instances in which oils might be preferred are for a prestigious one-off illustration, where time and expense are a lesser consideration; or for work principally carried out for private interest, when the artist may wish to exploit the rich qualities of oil paint and at the same time produce an original painting which is extremely durable — the colours and textures do not deteriorate over time.

Techniques
Underpainting and glazing are the main elements of traditional oil painting technique: dark shadows and mid-tones are laid with fairly neutral

tones and allowed to dry, at least superficially, before thin colour layers are overlaid, using paint thinned with turpentine and other diluents. Sometimes a mid-toned layer of colour is laid in first, and the underpainting includes both shadows and highlights, subsequently worked up in colour.

The alternative to this is to work with solid colour — the technique known as *alla prima* ('at first stroke'), blocking in an overall impression of the subject and gradually refining the detail. The expression 'wet into wet' means continuing to work the surface of the painting without allowing regular periods for the paint to start drying out. This means that the image is continuously manipulable, but it can also produce a muddied effect if the artist is unconfident of the medium.

Prepared canvas boards provide a ready-to-use surface for oil painting, although many artists prefer to prime a stretched canvas or wood panel to their own requirements. The traditional form of preparation is a coat of size, to seal the surface, followed by two or more coats of white primer. Each coat must be allowed to dry before the next is applied. Without

good preparation, the oils in the paint soak into the support and eventually cause deterioration.

MARKERS
Marker rendering has become a major graphic skill in recent years and is particularly suited to illustration of technical subjects, as certain textures which can be quickly achieved using markers lend themselves to simulation of metal, plastics and glass. There are obvious advantages to working with markers: the single medium combines attributes of both drawing and painting media, since the marker tip can be used to produce fine lines or broad areas of colour; and the technique is rapid and encourages a bold approach which gives a powerful impression of three-dimensional form.

The technique is economical and the effect of a confidently-drawn marker rendering is undeniably impressive. Herein lies the slight danger of the process, since it becomes easy to rely on relatively superficial interpretation and sleight-of-hand. For basic visualization of a product or an overall impression of a technical object, markers are the ideal medium. But many of the most authentically descriptive marker renderings show, on closer inspection, a degree of additional detail worked by other techniques which allow finer control of the finishing touches. Coloured pencil, ink or pastel are often applied to sharpen the fluid lines of the marker image. Hand-painted detail may be added using watercolour or gouache.

Techniques

Marker ink dries quickly and the passage of the nib leaves a hard edge to the line of colour. This produces the characteristically streaked or striped effect frequently exploited in marker work to imitate a shiny or translucent surface. It is possible, however, to create flat colour blocks: the smaller the area, the easier this is. The technique requires quick work to build up the colour spread while the leading edge is still wet enough to blend smoothly as the marker travels back to form the next line of colour.

Soft blending of colours is not a typical marker effect. The less absorbent the paper, the more subtle the transition of colour, because the ink remains wet longer on the surface. However, markers are available in a very wide range of colours, and an effect of gradation can be achieved

by working colour changes quickly one into another: for example, orange-red, followed by scarlet, followed by red-brown. Because the marker ink is translucent, it is also possible to form a gradation of tone by reworking progressively smaller areas of colour, creating a more dense hue in the areas which have been overworked. Allow the ink to dry

Above: Disc-wheeled bicycle
Artist: Anthony Lo
Marker work is ideal for presentation drawings.

Below: Video camera lens
Artist: Antony Papaloizou
Studio: Hop
Textural detail drawn with felt-tip pens is overlaid with marker work.

between applications to achieve a strong colour effect.

Impressions of colour mixes can be achieved by laying one colour over another. The same technique can also be used to build tonal values, working from dark to light — for example, shadows in black, mid-tones in warm grey, pale tones in a lighter grey, leaving white areas of paper to stand for the highlights. When overworking colours, check first whether the marker ink tends to bleed, and allow some drying time so that the technique does not result in a muddy, streaked effect, with no colour showing true.

The other problem to watch out for when using markers is the tendency of the ink to spread, travelling over the lines of a drawing into areas where the particular colour is not meant to be. This can be controlled by working just short of the outline, so that if the ink does spread it remains contained within the shape. When drawing fine lines with the narrow edge of the marker tip, any hesitation may form a blob of colour as the ink begins to be absorbed into the paper; in drawing both lines and blocks of colour, a fluid, vigorous approach is required, allowing the hand and arm to move freely across the drawing.

When using a plastic ruler or template to guide a marker, keep its edge lifted slightly from the surface to prevent an ink bleed. The ink does not dry as quickly on plastic, so wipe down the edge of the template before applying another colour, otherwise streaks of the previously used colour may appear in the new application. Low-tack masking tape or masking film, as used in airbrushing, can also be employed to define areas of colour application and create hard-edged shapes.

Right: Illustration for 'Into the Future' poster
© Ford Motor Company Ltd
Marker colour can be laid in rapidly sketched gestural strokes which add an appropriate sense of movement to an illustration showing new racing car designs.

APPLYING MARKER COLOUR

Flat colour **1 Apply even strokes, keeping the leading edge wet. Masking film helps to define the colour area.**

2 Extend the colour, working from side to side and allowing each stroke to merge with the previous one.

Blended colour **1 Apply the first colour as a graduated tone lightening towards the second colour area.**

2 Apply the second colour in the same way, reversing the direction, work over the graduated first colour.

GRADUATED COLOUR USING MARKERS

1 Work the lightest colours first in clean single strokes, leaving white areas for highlighting.

2 Build up the tonal range by overlaying mid-tones in banded layers, followed by dark tones.

Below: Car interior
© Ford Motor Company Ltd
Colour application is carefully controlled in this marker rendering to produce a realistic impression of colour and texture within the car interior. Roughly stippled textures overlaid with graduated colour correspond to the surface textures of the seats and overall fabric of the interior. Highlights are applied with opaque white gouache over the marker colours.

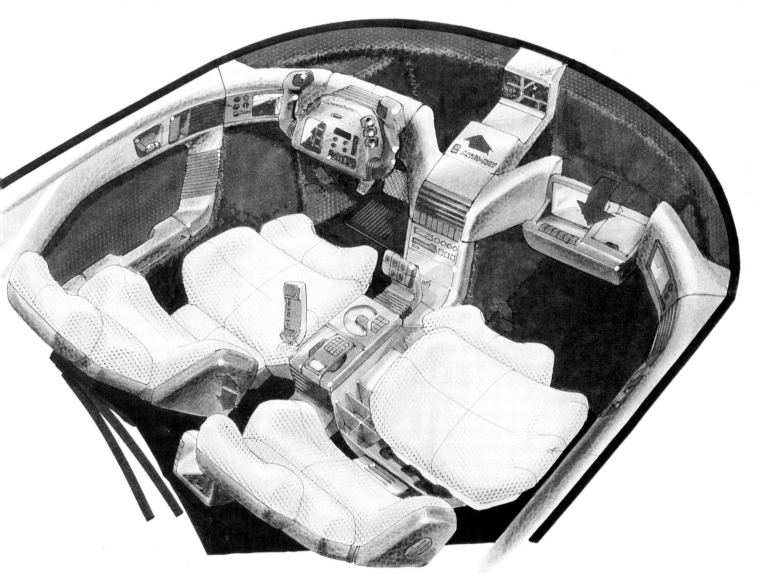

PASTEL

Pastel is, in effect, gouache in stick form, but the crumbly texture of the medium makes it better suited to a loose style of illustration rather than to precision and a high degree of realism. By frequent spraying of the developing image with fixative, pastels can be layered and over-worked as solid colour areas or broadly expressive linear marks.

Techniques

Quite crisp, fine lines can be made with the section edge of the pastel, while a broad, chalky line comes by applying the width of the pastel stick. A residue of loose powder colour spreads continuously across the surface, however, and it can be difficult to control the effects of line or colour. Some of the loose colour can be blown away or dusted off lightly using a paper towel or soft brush, but pressure on the surface settles the powder and dirties the lighter colours.

Pastel colour can be spread with a brush wetted with clean water, to create smooth colour blends. Another technique for laying a finely graded tone with pastel is to scrape off powder colour from the stick, using a knife blade, pick up the powder on a piece of cotton wool and rub it over the drawing surface. A thin layer of colour can be laid on the paper or

board surface and the deeper tones developed by repeated applications, softening the gradation by blending the colour layers lightly with the cotton wool.

A technique which eliminates the problem of laying broad areas of colour with pastel is to work on a tinted paper which sets the middle tone,

developing the darker and lighter values through the pastel work. This is best suited to renderings devised in a limited colour range, with a strong tonal emphasis.

Light pastel tints are often used for effects of highlighting on watercolour or marker renderings. If there is sufficient pastel detail to require fixing,

▶ USING POWDERED PASTEL

1 Use the blade of a scalpel or craft knife to scrape fine powder colour from the pastel stick.

2 Make a loose ball of cotton wool and pick up an even spread of powder colour on the lower side.

3 Rub the colour on the paper to form an even tone. This combines well with pencil, pastel or marker.

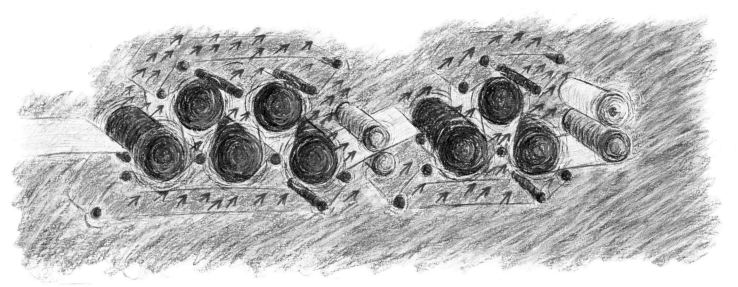

Above: Tough and technical: engineered fabrics
Left: Expanding in packaging
Artist: Zafer Baran, The Organisation
Design: Tor Petterson & Partners for Scapa Group
These two pastel drawings are from a series of fourteen commissioned for a company brochure to explain different facets of the client's business interests. The engagingly simple imagery and loose style of drawing is unusual in technical illustration, but these are employed to be illustrative of the technological aspects of packaging and fabric manufacture. The arrows are a theme running through all of the illustrations, creating a visual link but also serving a functional or descriptive purpose within each given context.

Pastel line **The section edge can be used to form fine sharp lines or sideways to lay broad, soft strokes.**

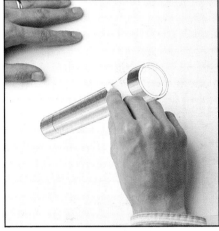

Highlighting **Pastel is a useful medium for laying highlights of pale, opaque colour over marker work.**

Right: **An effective method of working with pastel is to draw on coloured paper, allowing this to set the middle tone and keying the colour work to provide the lighter and darker ends of the tonal range.**

test the effect of a fixative on a colour sample of the paint or marker before applying this technique to a finished rendering. Pencil and pastel is an interesting combination of drawing media which provides a wide range of linear effects.

COLOURED PENCILS

Ordinary coloured pencils are designed to make linear marks. Even when used for solid-colour shading, the shape and width of the coloured lead creates a texture; completely flat colour is not a characteristic of coloured pencil drawings.

Techniques

The basic shading technique is to hold the pencil at an angle and use the side of the pointed lead to lay the colour as evenly as possible, rubbing across the paper surface from side to side. Colours can be blended together in this way, by grading off one colour and starting the application of the next over the faded edge, or overlaid to build up colour mixtures — simply, red over yellow produces an orange hue rather than the original pure red.

An alternative method of creating blocks of colour is to use the principle of optical mixing, applying short strokes of one colour, laying a second colour into the first in the same way,

Above: Ford Sierra
© Ford Motor Company Ltd
Markers have been applied to tinted paper to build the basic colour blocks. Over this, coloured pencils have been used to elaborate the form, the hatched textures counterpointing light and dark tones.

and so on, to produce an effect which at a distance merges into the required colour mixture. Overlaid hatching and cross-hatching — distinct lines rather than dashes of colour — produce a similar colour effect but with a different texture at close range.

Water-soluble pencils are initially applied in the same way as ordinary colour pencils, but the colour can be spread and blended using a paint-brush wetted with clean water. This allows the artist to combine effects of precise linear drawing with the fluid quality of watercolour painting, but using only the one medium.

Coloured pencil marks can be layered in different ways to enliven a surface or emphasize the detail in an image. After applying colour shading, for example, the point of a coloured lead can be used to draw linear detail over the colour blocks, or to scribble loose shadow effects. In an image of limited colour range, where tonal values are more important to the three-dimensional effect, a tinted paper can be used which establishes a mid-tone, and the details of highlight and shadow can be worked with light and dark coloured pencils.

COLOURED PENCIL SHADING

1 For block shading, work across the surface in a single direction with the pencil held at a low angle.

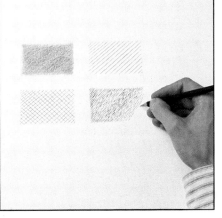

2 Colour blocks can also be built from ruled hatched or cross-hatched straight lines or with loose slashes.

This also saves time as there is no need to lay in such broad colour areas with the pencil.

This medium requires a degree of precision. It is difficult to erase the colour cleanly, and any damage to the paper surface tends to show up even after the colour has been reworked. A faint graphite pencil drawing can be used as the basis of the rendering. The sequence of colour application and the overall effects which it aims to achieve should be carefully thought out beforehand, although allowing that happy accidents can sometimes occur, as well as unwanted errors.

Below: Car seats
© Document Renault
In this rendering coloured pencils are again combined with marker work to give texture and depth to the colour areas, in loose stippling and shading which indicate the form and fabric of the car seats.

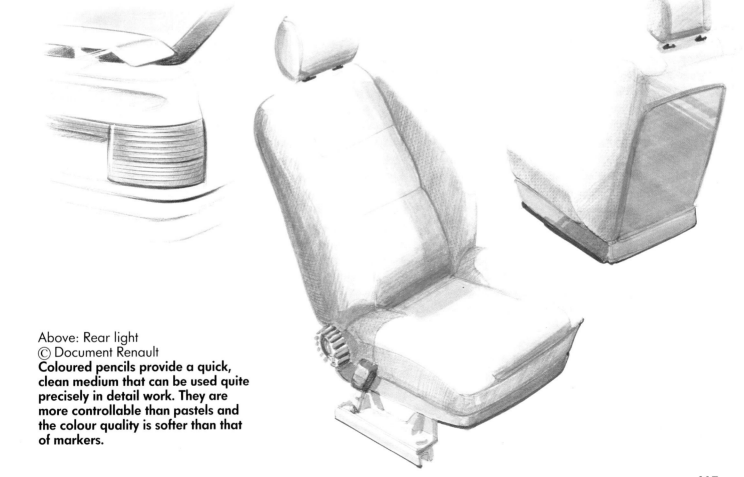

Above: Rear light
© Document Renault
Coloured pencils provide a quick, clean medium that can be used quite precisely in detail work. They are more controllable than pastels and the colour quality is softer than that of markers.

AIRBRUSH RENDERING

The airbrush is the only painting tool which lays colour without touching the surface of the support, and this is the key to its unique surface effects in colour rendering. The fine spray of colour that the airbrush emits can be used to grade colours and tones very subtly, building the layers of spray to a flawless finish. This produces the almost photographic realism exploited in many forms of airbrush painting, and particularly appropriate to certain areas of technical illustration.

Extensive use of airbrushing to produce original colour rendering is a relatively recent development in all fields of illustration, although the airbrush has been used in technical illustration for some time as a means of applying subtle coloration to complex imagery. For many years, however, airbrushing was an important technique in photo-retouching and was more commonly exercised within this area of graphic imagery. Black-and-white bromide prints were retouched using a specially developed range of grey pigments comparable to photographic greys. Retouching enabled artists to 'clean up' pictures to be used in catalogues and publicity material: for example, taking a shot of a working machine in context and brushing out any blemishes on the machinery along with the background area, to leave a clean image on a plain ground. The airbrush was, and still is, also used for colour retouching in certain circumstances, although photographic techniques have been so highly developed that a great deal can now be done to manipulate an image through developing and printing processes.

AIRBRUSH CONTROL
To create the precision and detail of a highly rendered technical illustration, the airbrush artist has to rely on masking techniques used to separate areas of different colour and give crisp,

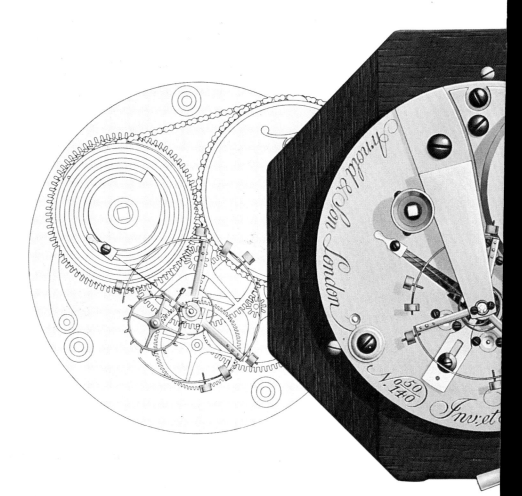

CONTROLLING THE AIRBRUSH SPRAY

The width of the colour spray from the airbrush partly depends upon the height of the tool above the surface.

By bringing the airbrush in closer to the surface, the artists has tighter control of the colour flow.

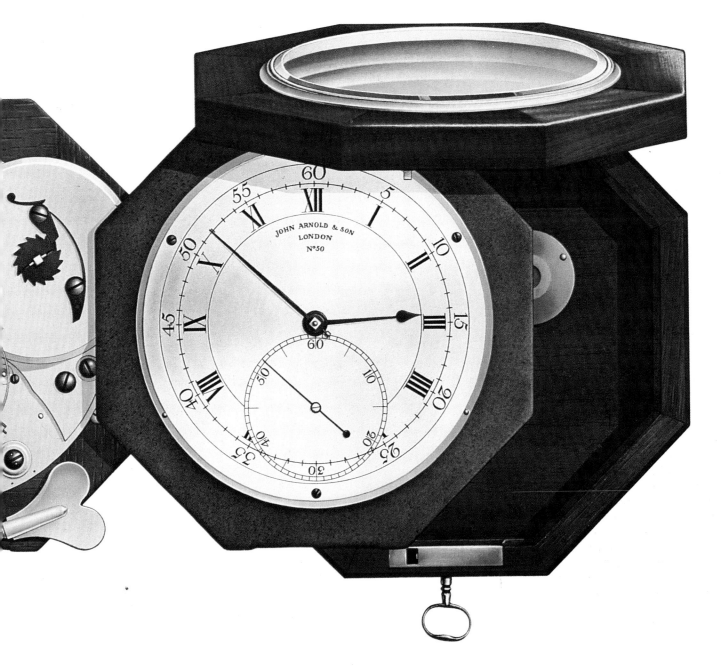

Clock
Artist: David Penney
A skilled airbrush artist can produce an effect of extraordinary realism by careful control of tonal gradations and colour blending. It is often necessary, however, to work the final details by hand when the process of masking and spraying cannot achieve the delicacy required. This illustration contrives a balanced and interesting combination of fully rendered representation and diagrammatic line work.

hard edges to the shapes. A certain amount of freehand control is always required, however, in modelling the areas of graduated light and shade within the masked shapes, and in simulating surface textures.

The newcomer to airbrushing should spend some time growing accustomed to the push-button control of the airbrush and learning to adjust the spray pattern both by varying the flow of medium through the airbrush and by raising and lowering the instrument above the support. The closer to the surface the airbrush noz-

zle is held, the narrower the 'cone' of spray. Working closely in this way, the airbrush artist can apply fine lines and tiny dots of colour. As the airbrush is pulled away from the surface, the spray area increases, eventually giving a broad spread of colour. The artist also needs to gain a smooth action in starting to spray and stopping when the work is complete, as an abrupt motion translates into a variation in the width and density of the spray.

Simple exercises consist of spraying single lines, gradually varying the width, followed by evenly worked

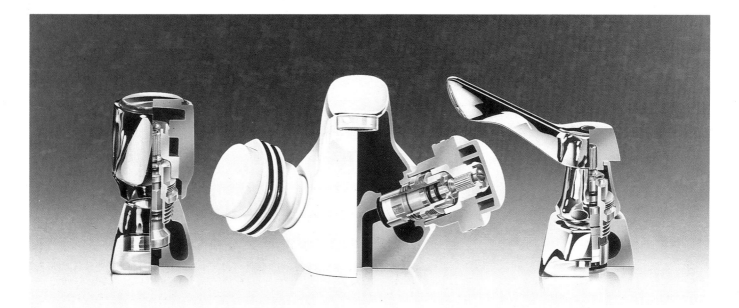

criss-crossed lines; and a consistent pattern of dots, regularly spaced and equal in size, then gradually increasing in size. It is useful to practise spraying lines both freehand and by guiding the airbrush along the edge of a ruler. The ruler should be tilted up from the surface to maintain the airbrush nozzle at the required height for the width of line.

GRADUATED TONE AND COLOUR

Imperceptible degrees of shading in the transition from light to dark and a similarly subtle merging of colours through the range of the spectrum are the characteristic effects of airbrushing. These techniques are used to model three-dimensional form and to simulate particular surface qualities

Faucets
Artist: John Brettoner, Aircraft
Design: Edward Briscoe Associates
Agency: John Bowler Associates
Client: Barking Grohe
Airbrushing creates subtle tonal gradation. Masks can be used to obtain hard-edged contrasts for rendering of smooth-surfaced or intensely reflective materials.

▶ SPRAYING GRADUATED TONE

1 Mask off the area to be sprayed. Start spraying across the top of the masked shape, allowing the colour to overlap the mask edges.

2 Move the airbrush evenly from side to side and work downwards, gradually drawing away from the surface to lighten the colour density.

3 When a smooth gradation of tone has been achieved, allow the colour to dry and remove the mask carefully.

Above: **The fine quality of airbrush spray enables the artist to build subtle gradations of colour and tone. The transparency of watercolour (above) allows a smoother transition from one colour to another, as the colours blend on the surface. Using gouache (top) it may be necessary to mix the intermediary colours separately.**

Right: Shower pump
Artist: John Brettoner, Aircraft
Design: Edward Briscoe Associates
Agency: John Bowler Associates
Client: Barking Grohe
A special technique was used to achieve the bronzed metallic effect in this image. Colour was rolled onto a gesso base using a small sponge to produce an initial texture. This was oversprayed with transparent colour, keeping the airbrush at a very low angle, and a hard eraser was then used to rub back the sprayed surface to create the highlights.

— the soft sheen of smooth plastic, for example, or the more intense effect of a shiny metal.

Overall tones are achieved by layering successive applications of spray. This is an important aspect of the technique which maintains the smoothness of the surface effect. If colour is laid too heavily in a single application the surface may become wetted out and, especially on a smooth artboard which does not 'grip' or absorb the medium, there will be variations in the surface quality and the colour will dry unevenly.

A flat tone is achieved by spraying evenly across the image area from side to side, gradually moving down the shape, then repeating this action until the required colour density is created. To form a graduated tone, a similar technique is used but the airbrush is steadily pulled further from the surface to grade the colour evenly into the lighter area of tone. The darker section requires more respraying to build the tonal density, so a smaller area is treated in each successive spraying. It is important to keep the layers of spray light and even to avoid creating a striped effect of abrupt transition from dark to light.

Merging two colours is easiest with a transparent medium such as ink or watercolour, since the central area where the two colours merge actually creates a colour mix through the layering of successive applications of colour. With the opaque medium of gouache a smoother effect is achieved in merging two colours directly related in the colour spectrum (such as yellow and orange) than two opposite colours (such as orange and blue). To create a smooth gradation through contrasting colours, it is necessary to mix the intermediate hues and spray them separately.

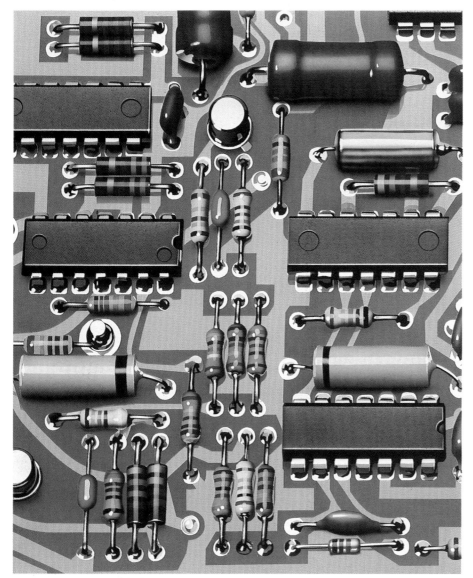

Left: Circuit board
Artist: John Harwood, Folio
Hard-edged highlights give the glossy appearance to the brightly coloured elements on the circuit board, but these are interwoven with subtle gradations describing the cylindrical forms. The illustration presents an excitingly abstract image.

MASKING METHODS

For precisely defined, hard-edged shapes on a large or small scale, the illustrator uses masking film to mask off the outlines as each area of the image is coloured. The purpose of a mask is to protect areas of the surface which are not to be coloured at any given stage; this includes unsprayed areas and parts of the image that have already been airbrushed. Because the plastic masking film is adhesive, it grips the support, completely sealing off all but the area to be sprayed. Within a film-masked area, the airbrush spraying can be manipulated to create soft gradations of tone or linear textures, so the masking process does not necessarily limit the image to separately defined areas of colour. However, a flat, highly graphic style is also readily achieved with airbrushing, by spraying transparent colour over line work using masking film to maintain the precision of the image.

The sequence of masking for a particular image needs some forethought. It depends upon the complexity of the image, the medium used for the airbrushing, and the individual artist's preferred method of work. A straightforward representation worked in watercolour or ink may require only one application of masking film. The entire image area is covered with film, allowing an outer border all around. The image is worked from dark to light: that is, the masking film is cut and removed from the darkest shadow areas first. Subsequently mid-tones are sprayed, moving on to the lightest tones last. Because the medium is transparent, the gradual layering of the spray only increases the tonal density quite subtly in the darker areas, so these do not need to be remasked before the mid-

APPLYING MASKING FILM

1 Peel the backing paper from the film along one edge and smooth the film down on the artboard.

2 Rub down the film across the full width, gradually extending the area in contact with the artboard surface.

Above: **Loose masks produce a softer effect than film masking and give variation in edge qualities. Cut and torn paper or card can be used. A linear texture (top left) is achieved by spraying through the teeth of a comb.**

and light tones are overlaid.

When working with an opaque medium, it is possible to spray dark over light. The masking method here may involve working up small sections of the image separately, remasking each section before working the next, and then adjusting the tonal balance overall when the image is complete.

Masking film is cut in place, using a lightweight scalpel (utility knife), taking care not to score the underlying surface. It is usual to keep cutting on the artwork surface to a minimum, avoiding continual application of new masks. Some artists prefer to mask and cut the whole image as the first stage, then lifting mask sections in sequence as the work progresses and replacing them as necessary. But the gradual build-up of sprayed colour on the surface of the film eventually disguises the previously sprayed sections of the image below, making it difficult to balance or match colours and tones. Masking film can be lifted partially at various stages of the work to check the overall balance of newly

applied colours and previously sprayed areas, but sometimes complete remasking is required.

Sections of masking film that have been cut away are usually saved for replacement by storing them on the original backing sheet of the film. This keeps the adhesive side clean. As the film is soft and can stretch, it is sometimes difficult to replace mask pieces accurately. Paper masks and transparent, low-tack tape can be used in conjunction with masking film

Above: Aircraft cockpit
Artist: Ian Stephen, Spectron Artists
Client: Marshall Cavendish
The slightly diffused quality which can be obtained with an airbrushed line gives a fluorescent glow to the silhouettes of the figures and similarly imparts a dazzling effect to the runway lights.

Underwater suit
Artist: Nick Hawken
Highly skilled airbrushing technique and rigorous analysis of surface effects in transparent and reflective materials are needed to produce such flawless photographic realism.

to mask off broad areas, fine details and straight edges.

Loose masking

Since masking film adheres to the surface, it creates a clearly defined, hard edge to the area of sprayed colour. Softer effects can be achieved using loose masks, made of paper, card or acetate. These are held on or above the painting surface: the further from the surface the edge of the mask, the softer the edge of the sprayed area, because the fine particles of colour drift slightly as they pass under the mask. A cut paper mask also makes a different effect from that of a torn edge — this is an area worth exploring. Loose masking is used in conjunction with film masks, to create a more freely worked effect within a hard-edged colour area, or to add textural detail in broad areas of colour and tone.

Distinctive textures can be obtained by using other materials, or even solid objects, as loose masks. For example, spraying through a loose-weave fabric creates a rough but regular texture; spraying through the teeth of a comb provides a loose linear pattern. As much technical illustration depends upon precision in the drawing and definition, these techniques are not always appropriate, but can be used to insert contained areas of texture within a shape strictly defined by masking film. In some types of editorial or advertising illustration, however, there may be a broad depiction of a background beyond the focused object which provides scope for increasing variation of airbrushing technique.

Liquid masking

For areas of small detail or overall texture, liquid masking can be painted into the image with a brush, or sprayed with the airbrush. This method can be used to develop detail within an area which is already masked with film or paper but is not suitable for broad areas of the image. The liquid dries to a rubber skin which seals the surface and colour can be applied over it; the rubbery masking is lifted when the colour is completely dry by peeling it back gently with the tip of a scalpel (utility knife) blade, or by rubbing with an eraser. If masking fluid is sprayed through the airbrush, the tool must be flushed out thoroughly with water as soon as the masking is completed.

SHAPE AND VOLUME

Masking film allows the artist to identify precisely the outline or contour of a form, but all it does is to delineate a flat shape on the artboard. Details of three-dimensional depth, angled and curving planes, solid or hollowed shapes have to be defined by the airbrush work. This is a matter of modelling the range of tones accurately to establish the three-dimensional form unambiguously. Volume is described by a pattern of light and shade, and the basis of the artist's interpretation depends, once again, on accurate observation of the real form and an ability to translate that into a logical system of representation.

Many technical illustrations look

enormously complicated, but this is usually to do with the amount of detail incorporated in one image, not the complexity of any individual form. The image is built up from a number of simple elements which tend to recur; these are combined in different ways to produce compound forms. The most basic elements of airbrush representation are flat and curved planes. These can be initially investigated through basic exercises in rendering a few simple geometric solids — cube, cylinder, sphere, cone — which present flat and curving planes in different relationships to one another. The cube is a system of flat planes set at regular angles; the sphere is a continuous curve in all directions; the cylinder combines a continuous curve in one direction with a flat circular plane through the cross-section of the form; the cone functions similarly to the cylinder in surface effect, in terms of airbrush technique, but with the regular diminution of the horizontal cross-section leading to the point of the cone.

Learning to render these simple forms in solid volume represents a useful exercise in controlling the application of graduated tone to a specific shape. The cylinder, for example, relates very directly to the type of problem which may be confronted in a full-scale illustration, as so many technological subjects have cylindrical components — shafts, pipes, wires — and elements based on cylindrical form, such as springs and screw threads. Wheels, gearing cogs and fan mechanisms often correspond to a basic form which is a 'slice' out of the cylindrical cross-section. So it is frequently possible to relate a new problem of representation to a principle which you have already understood and mastered.

In addition, however, the illustrator is confronted with irregular and complex forms which are not directly translatable in these terms. The varying curves of a car body or an engine casing, for example, are not necessarily related to a symmetrical geometric form. If there is plenty of photographic reference which identifies the pattern of light and shade describing the surface contours and underlying volume, this can be directly interpreted as a tonal scale for the airbrush work. Sometimes, however, the illustrator has to assume a source of illumination and its effect on the surface modelling of the object. In this case, it is helpful to define the general condition of a given area of the object: curving away in all directions as in the sphere; forming angular flat

Airbrushed exercises
Artist: Kevin Barnes
It takes practice in masking and spraying techniques to acquire the skill to render convincingly different effects of volume and surface texture. Exercises such as these which examine methods of developing specific forms and surface qualities with airbrushed colour are valuable groundwork for more complex projects.

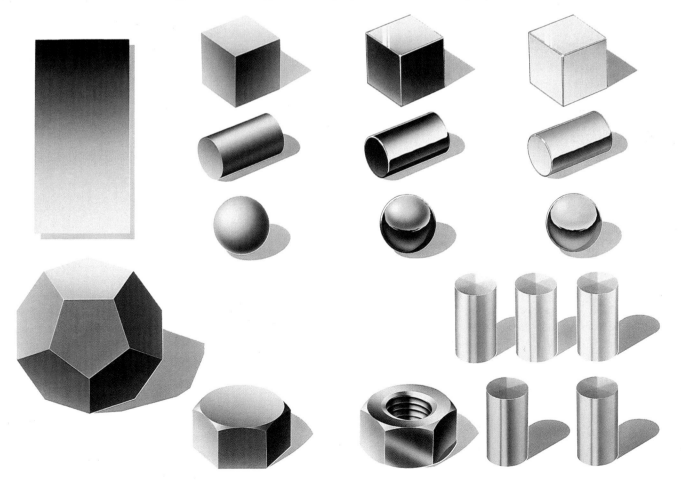

planes as in the cube; a directional curve as in the cylinder; flaring out from an apex as in the cone. This type of analysis becomes an unconscious habit as the airbrush artist gains more experience of colour rendering and of the capabilities of the airbrush.

HIGHLIGHTING

The distribution of tonal values — shadows describing volume, cast shadows throwing forms into relief, highlight areas representing maximum illumination, and the range of mid-tones linking the extremes — are analysed and interpreted in the same way in airbrushing as in tonal drawing or in any other form of colour rendering. What has become a hallmark of highly finished airbrush art, particularly in subjects dealing with man-made forms and materials, is the additional degree of surface polish which can be achieved by specific airbrushing techniques and related effects worked by hand over airbrushed colour.

Methods of highlighting can be exploited very subtly or with bold panache in airbrushing to give an image extra crispness, soft luminosity or exuberant sparkle. A highlight is simply defined as the lightest area in an image. In airbrush work using a transparent medium, a significant amount of highlighting is created simply by leaving areas of the artboard white, grading in the colour softly to maintain surface coherence. In gouache painting, a light veil or solid burst of opaque white is used to highlight a form broadly or finely. Otherwise, special effects of highlighting are obtained by different methods to concentrate the illumination to an edge or high point of a form, increasing the visual attraction of the image. This is especially important in, for example, images used for product promotion in sales and advertising contexts.

The airbrush itself can be used to spray focused dots and slashes of diffused light, from a pinpoint highlight to the spectacular airbrushed starburst commonly used to add dazzle to highly reflective materials such as chrome or glass. These are applied

with opaque white over previously airbrushed colour. The airbrush spray provides a faint diffusion of colour at the edges of the highlight, suggesting intense light radiation. More solid effects can be applied as line or dot highlights by painting with a fine sable brush or applying opaque white with a ruling pen. These are additive methods of highlighting.

Alternative forms of highlighting involve removing colour in selected areas. One method of doing this is to scratch back the layer of spray using the tip of a scalpel (utility knife),

revealing the whiteness of the support through the surrounding colour. This method provides a variety of effects, from minutely fine lines to broader areas of solid white. It should be used very selectively: you need a light touch to avoid damage to the underlying surface, as the cleanness of the artboard showing through is essential to a successful result. A different effect is achieved by using a pencil-type eraser — either given a fine point or used as a blunted tip — to rub away the colour gently. Both of these methods allow the artist to 'draw' into

► STARBURST HIGHLIGHTS

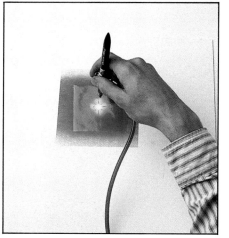

1 Cut the shape of a star in the centre of a piece of acetate. Spray with opaque white on a dark ground.

2 This method produces a hard-edged shape. Intersecting lines cut in the acetate produce a fine linear star.

3 To create an effect of radiating, diffused light, spray a soft burst of white over the centre of the star.

4 This produces a softly haloed effect. The basic principle can be adapted to linear or multiple starbursts.

the image after the colour has been applied and produce some very interesting textural effects, suggesting varied sources of illumination and qualities of surface reflection.

SURFACE TEXTURE
The general requirement for accurate representation in technical illustration presents the artist with particular problems in rendering convincingly the surface textures of a variety of man-made and natural materials: the range of metals used in construction work and precision machinery; plastics of varying quality with matt,

ERASED HIGHLIGHTS

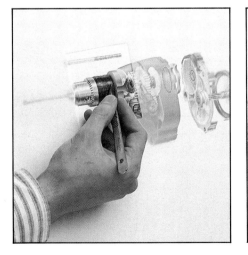

Use a scalpel to scratch back finely detailed highlight areas. Protect the artwork with a paper 'window'.

The tip of a hard pencil-type eraser rubs the surface colour away to produce soft-edged highlighting.

Above: Corvette
Artist: Vincent Wakerley
Studio: Visage Design
Strong tonal contrasts and dazzlingly bright highlights are cleverly distributed to represent the surface gloss of paintwork, chrome and glass.

shiny or textured finishes; rubber; wood; stone; brick; glass and transparent plastics. Primarily, this is a question of identifying the surface characteristics of the material accurately, to translate the particular qualities into tonal and colour values which can be easily interpreted with the airbrush.

The process of layering tone and colour in airbrushing helps the artist to reproduce subtle differences of texture. Broadly speaking, matt-surfaced materials require a soft-edged

Right: Cutaway tyre
Artist: Kevin Jones
Client: Michelin Tyre plc
Subtle tones of airbrushed colour in watercolour and gouache are combined with line work and hand-painted detail in this layered cutaway.

approach, using gentle tonal gradation and possibly a combination of loose and hard masking, while a shiny surface presents hard-edged shapes and a greater contrast of tonal values.

Sometimes a textured masking material can be found which corresponds very directly to the texture required in the representation. The airbrush itself offers a particular textured effect known as spattering, corresponding to stipple techniques in drawing. This consists of a coarser pattern of spray giving the effect of myriad dots of colour. Some types of airbrush can be fitted with a spatter cap which automatically produces a coarse but fairly even spray. Alternatively, the effect can be achieved with any independent double-action airbrush by lowering the proportion of paint to air, disrupting the evenness of the spray pattern.

The subtractive techniques of highlighting previously described — scratching back with a blade or rubbing away with an eraser — increase the textural range. These techniques do not have to be used only for highlights: the scratched or rubbed area can be resprayed with transparent colour to convert the highlight into a surface pattern or texture integrated with the overall tonal values of the rendering.

AIRBRUSHED IMAGES
Airbrushing has become the dominant technique in technical illustration for producing highly finished, polished colour renderings with a photographic authenticity. One advantage of airbrush work is that it can apply this degree of 'super-realism' to a view of the object which could not be directly achieved by photography. This is exploited particularly in cutaway and ghosted views (examined in detail on pages 138–153), forms of representation which allow the viewer to see into and through the structure of

AIRBRUSHING TRANSPARENCY

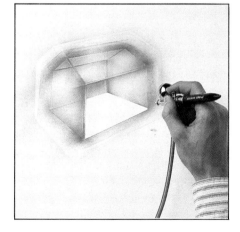

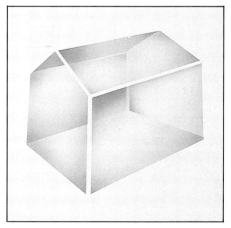

1 Using film masking, spray every facet of the object in turn with transparent watercolour or ink.

2 When the spraying is complete, remove all remaining mask sections to reveal the clean-edged image.

3 In a transparent effect, the tonal gradation defines the depth between background and foreground planes.

an object while maintaining a sense of solidity and three-dimensional form. But the quality of surface finish achieved through airbrushing is closely associated with many forms of technical illustration: from flat colour evenly applied to line work, to the complex three-dimensional modelling of sophisticated images representing equally sophisticated objects.

There is a superficial glamour to the end result of airbrushing which sometimes disguises the amount of methodical preparation that has gone into constructing the image before the airbrush is brought into play. Airbrushed illustrations depend upon skilful drawing as much as upon the skill of manipulating the airbrush; it is merely a tool available to the artist, though one which contributes a unique visual element.

Motorcycle race
Artist: Stuart McKay, Aircraft
Paper masks and spattered spray were used to obtain the soft-edges and hazy texture of this image.

EXPLODED VIEWS

The exploded view is a catalogue of parts presented graphically. It shows a whole object by isolating all its components, with each part identified separately but in correct relation to the other components and in the serial order of construction. The origin of this convention lies in the practical needs of engineering and manufacturing: the exploded view provides instruction for assembly or maintenance of a technical object. It has subsequently been adapted to other fields of illustration work, such as educational publishing or sales brochures, as a dramatic form of graphic presentation which is more deliberately explanatory than a straightforward single view.

THE LAYOUT OF AN EXPLODED VIEW

A typical example of an exploded view shows an object which is basically cylindrical so that the parts lock together along a single central axis; this might be, for example, a motor, a clarinet or a technical pen. The purpose of the illustration is to show every item contained within the construction, including internal components normally hidden, such as washers or O-rings which fit inside larger components, and interior elements such as screw threads.

The exploded view pulls the components apart and displays them 'threaded' along the central axis. The effect was at one time physically achieved in the practice of a particular industrial drawing studio, by disassembling the object and stringing the components on to a glass rod; when they were disposed along the rod at the required distances, photographs or drawings could be made providing a direct representation. In modern practice, the illustrator will, if possible, take the object apart to research the drawing; sometimes, however, the framework of the illustration has to be constructed from existing drawings of single parts and assembled views.

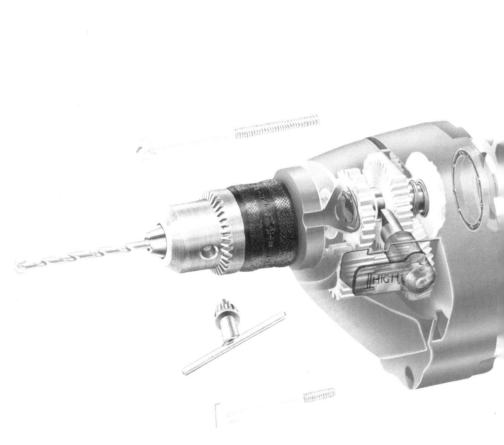

The sequence of assembly cannot be changed for purely visual or aesthetic considerations. Exploded views are the most accurate and unvariable form of technical illustration: the layout is absolutely dictated by the construction of the object. In parts manuals used for assembly and repair in an industrial context, each item is also numbered in sequence and named in a table of contents, but this is uncommon in editorial illustration unless the accompanying text explanation is very detailed and complex.

Another graphic device which may be used to explain the exploded order is to show the central axis as a dot–dash line; this passes through the components which are inked in solid line, leading the viewer's eye through the scheme of assembly. This convention has also been adapted to colour renderings, where sometimes a line of colour clearly distinguishable from other elements of the rendering travels through an exploded detail or the overall exploded view.

There is no standard format for the distance between components. They can be either completely separated or allowed to overlap. The spacing

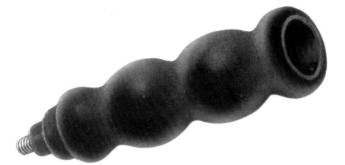

Power drill
Artist: Michael Gilbert
The basic conventions of an exploded view are applied to this detailed illustration of a power drill, based on a pencil drawing of the disassembled object. Additional information is supplied by introducing ghosted and cutaway detail in specific parts to show the complexity of the construction. The main colour areas were airbrushed in watercolour and gouache, with textures and highlighting developed by fine line brushwork.

Prototype car
Artist: Mathias Kulla
A development project for a new automobile is shown here as an exploded view to explain the relationship of parts to the lay viewer. Marker work is used for the presentation drawing to provide a lively, descriptive colour rendering of the original pencil sketch.

between components may partly depend upon the overall format of the illustration and how much space there is for laying out the exploded view vertically and laterally. It also depends upon the detail to be shown: for example, if there is a flat casing on the front of a component it may be necessary to pull it completely apart to reveal detail on the inner face; alternatively, there may be no such internal complexity, in which case parts may be shown slightly overlapping in the exploded view without losing any important information. For

work on technical manuals, there may be a house-style to follow, aimed at standardizing the forms of graphic presentation for practical interpretation by assembly workers and servicing staff. In editorial work, the illustrator will probably have complete discretion in constructing the layout of the exploded view, providing that the explanatory function of the illustration is clearly served.

Not every technical object is cylindrical, nor can the sequence of parts necessarily be encompassed by following a single direction through the image. The principle of exploding along an axis is nevertheless applicable to all types of constructions. A vertical or horizontal axis can be fixed around which the basic structure is balanced. In a complex item where there are various assemblies which interlock, the taking apart of the components follows this principle in each section, and the layout of the drawing will have a centre from which these sections line up or radiate in a logical relation. Thus, for example, an object

Car chassis
Artist: Hugh Dixon, Spectron Artists
Client: Citroën Cars Ltd
The exploding of parts in this rendering is very controlled. Sections of the car body are pulled out just enough to show how they interlock within the overall design and to reveal some normally concealed surfaces such as the inner edge of the door. Airbrushed colour creates a uniformly sleek surface effect.

which has a horizontally layered construction, such as a slim pocket calculator, can be exploded on a vertical axis. Something which is basically cubic, such as a washing machine, is exploded by pulling out each side on a direct line from the common centre. Fixing components such as screws or bolts which run through the cross-section of a main form may sit at 90° to the central axis of the drawing, on a minor axis passing through the centre of the screw and the centre of the hole into which it fits.

The more complicated the construction and the number of parts to be described, the more it becomes a basic design problem to lay out the exploded view in an unambiguously descriptive form. Returning to the example of the cylindrical object exploded on a central axis, if this object has, say, 40 components, the exploded view is, in simple terms, very long and narrow. How is the illustrator to fit this into the basically rectangular format represented by the paper or board on which the illustration is worked, or the book page on which it will be printed? Therefore, it is not always possible to present the axis as a continuous straight line, but if this line is cranked to form an S-shaped layout, or the overall view is broken into sections positioned side by side, some graphic means must be found of directing the viewer through the illustration in the correct sequence. The possible solutions to such a problem need to be worked out through thumbnail sketches and rough layouts before the artist can get down to the task of drawing the exploded view in detail.

VIEWPOINT

Where the exploded view is intended to have a strictly practical application, it may be presented systematically by means of an isometric or axonometric projection working from orthographic data (see pages 64–69). Without producing a visually realistic interpretation of the construction, this explains the three-dimensional relationships and pursues the strict logic of the assembly. It is basically a schematic solution.

A perspective view may be more difficult to work out, but it provides a directly recognizable presentation to the lay viewer, so this is commonly used in illustrations for sales, teaching aids, practical books and magazines for a general readership. If the purpose of the illustration is simply to clarify the physical make-up of the object, the perspective view is constructed in realistic terms according to the true scale and dimensions: for example, there is a minimal effect of perspective in a small portable object. If the illustration is designed mainly for visual impact, to draw the viewer's attention towards some particular aspect of the information supplied, a

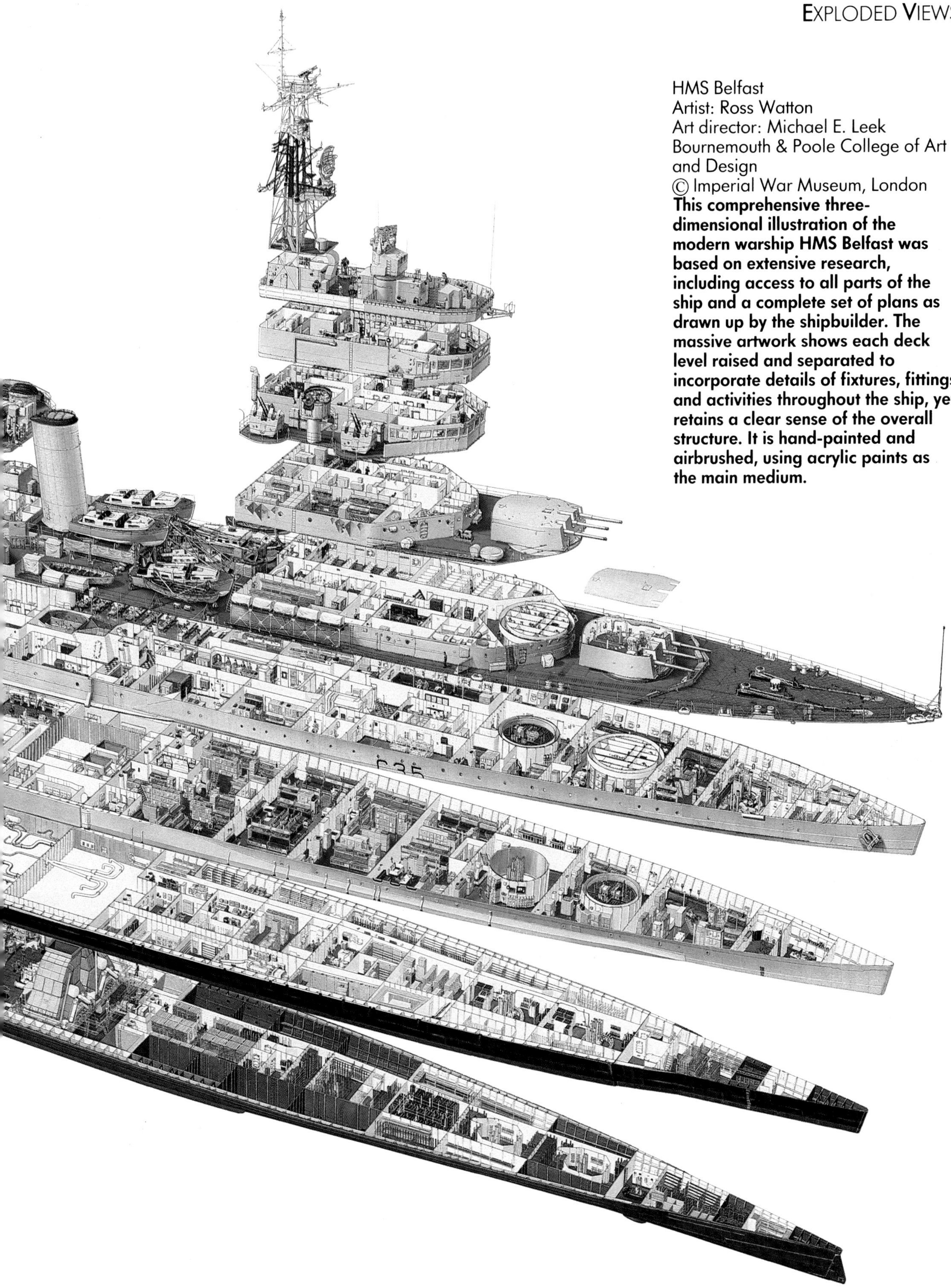

HMS Belfast
Artist: Ross Watton
Art director: Michael E. Leek
Bournemouth & Poole College of Art
and Design
© Imperial War Museum, London
This comprehensive three-dimensional illustration of the modern warship HMS Belfast was based on extensive research, including access to all parts of the ship and a complete set of plans as drawn up by the shipbuilder. The massive artwork shows each deck level raised and separated to incorporate details of fixtures, fittings and activities throughout the ship, yet retains a clear sense of the overall structure. It is hand-painted and airbrushed, using acrylic paints as the main medium.

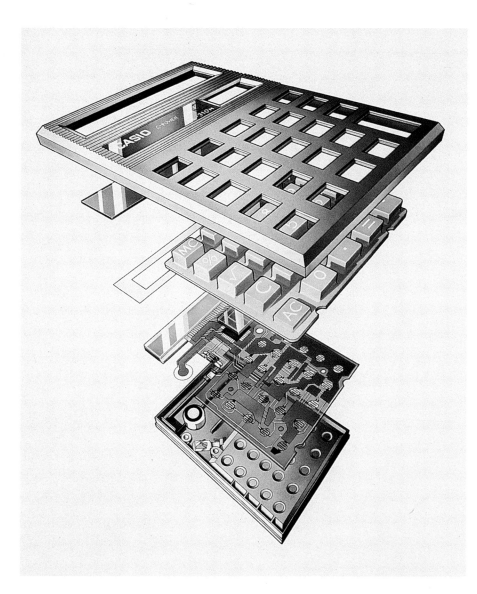

Pocket calculator
Artist: Sean Wilkinson
The slim form of the flatly-shaped calculator is exploded on a vertical axis with diminishing perspective giving it an almost epic scale. This makes a visually interesting design of the simple layered construction while displaying all the necessary information about component parts. The airbrush work emphasizes cast shadows and reflected highlights to enhance the three-dimensional effect.

THE WORK IN CONTEXT

Much of the information in a technical manual may be presented in the form of exploded views, and this is almost invariably done as line work, since graphic presentation values are of minimal importance compared to practicality, and cost-effectiveness in preparing and reproducing the illustration work is likely to be of some importance. The original illustrations, usually inked on drafting film, can also be easily updated to accommodate technological developments — a more sophisticated single component incorporated in a specific assembly, for example, or a modification designed to meet different control standards when a product is sold into an overseas market. The particular section of the drawing is erased and the new detail drawn in. For an illustrator working in this context, as much time may be spent in amending existing drawings as in producing new material.

In the area of general publishing, the exploded illustration is most likely to be a one-off production accompanying a practical text. The overall scale and the method of rendering are established in terms of the publication's visual style and fixed format. Line work may still be appropriate for reasons of clarity or economy, but the attractiveness of the presentation has greater significance. Colour may be introduced as simple tint patches added to the keyline drawing, or might extend to full colour rendering produced by hand-painting or airbrushing techniques.

relatively exaggerated perspective might be employed to dramatize the graphic effect.

SUPPLEMENTARY DETAIL

Standard visual devices for adding information to an exploded view include diagrammatic cross-sections relating to particular parts of the assembly and, in a highly complex rendering, enlarged details of interior components. This occurs when the varying scale of individual elements makes it impossible to plan a single exploded view which can encompass the full range while remaining graphically readable. Standard methods of presenting separate detail views include placing the enlargement within a circle or box, with a leader line pointing to its position on the main

view, or annotating the artwork and using a reference number or letter to key in the detail enlargement.

An alternative use of the exploded view is as a detail of a broader rendering, to focus a particular aspect of the subject and provide more detailed explanation. For example, accompanying an illustration of an airplane cockpit, there might be an enlarged-scale exploded view of the ejector seat assembly; for any motorized object, a visualization of the entire object can be supplemented with an exploded diagram of the actual motor mechanism. It is often necessary to combine illustration conventions to evolve an image which both describes a technical object and incorporates quite specific levels of information.

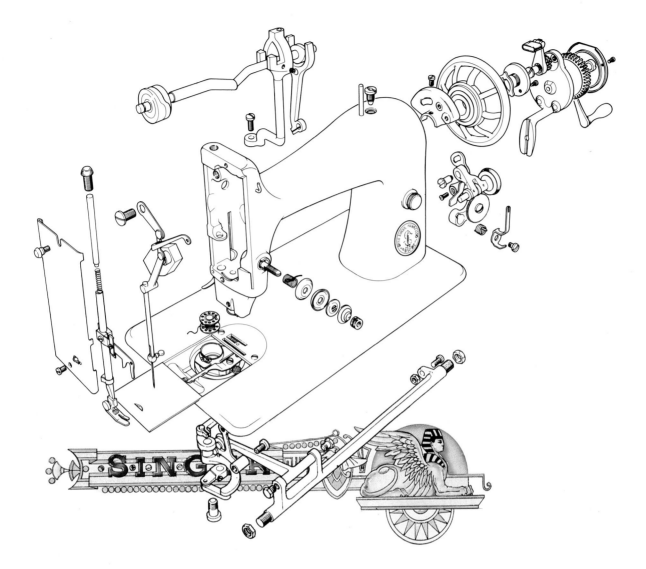

Above: Singer sewing machine
Artist: Emma Collins
Portsmouth College of Art, Design
and F.E.
**The sewing machine, in common with
many domestic technical objects, is
designed for practicality with a
smoothly formed outer shell which
houses and protects complex areas
of delicate machinery.**

Right: Sub-sea thruster
Artist: James Sunderland
**This is a well executed version of the
standard form of an exploded view
applied to a basically cylindrical
object. It includes a cranked axis line
linking the two sections of the
drawing, and line width variations
within the forms to distinguish inner
and outer angles and surfaces.**

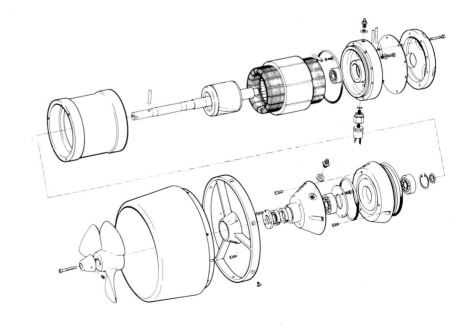

CUTAWAYS

The term cutaway refers to the convention in technical illustration of showing an object as if there was a piece taken out of one or more layers — by piercing a hole through the surface casing or slicing into the solid mass of the object — to show interior components and mechanisms. This might mean removing part of a wall to show machinery installed inside a building, or cutting through the outer shell of a boat to reveal its fittings; or on a smaller scale, 'breaking off' a corner of a radio to reveal the working parts or cutting a wedge-shaped segment into the cross-section of a cylindrical instrument to examine the constructional layers. The cutaway is used to analyse the workings of technological items in many different contexts. As a means of demonstrating the form, function and interrelationships of components it can be applied to opening up the giant bulk and complex systems network of a nuclear power station, or the simple and economical form of a reservoir pen. The cutaway is a device which works equally well in a linear, diagrammatic or full colour rendering.

SELECTING THE CUTAWAY VIEW

The section of the object treated as a cutaway depends upon the arrangement of interior parts and the amount of information the artist wishes to convey. By definition, some part or parts of the outer surface are shown through which the cut is made, with the exception of a straightforward cross-section through a profile view, when the cut edge is also the outline of the main image.

If the object is of regular construction, a single cutaway section may be representative of an interior structure that continues uniformly through the full length or width of the object. Alternatively, one cutaway area may open up a broad or deep view that reveals various aspects of a complex interior. If the intention is to focus on a single important feature, the illustration must

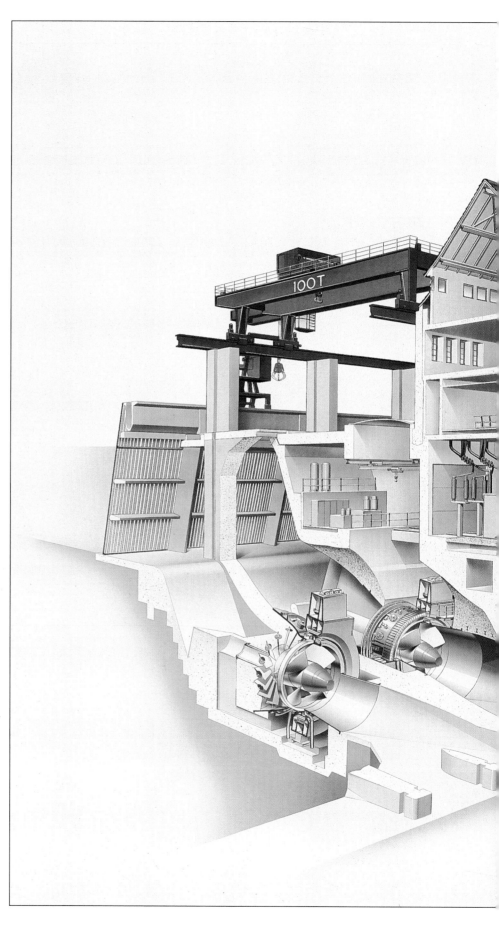

Hydroelectric plant
Studio: Warwickshire Illustrations Ltd
Working structures on architectural scale lend themselves to cutaway presentation. The cut sections in this illustration are indicated by a stippled texture on the cut edge of brick or concrete elements, with the conventional flame red line applied to cuts through metalwork and machine housings. The three-quarter view, sliced through at right angles to the outer wall, allows the cutaway areas to be seen in relation to a detailed description of the exterior.

be highly selective in positioning the cutaway view. However, if the idea is to represent a complex of different components and functions, it may be necessary to create separate cutaway areas explaining individual parts, or to organize a framework containing as many cut sections as are necessary, while maintaining the distinction between separate elements. When a cutaway view is used in illustrating a large-scale construction, the cut edge may frame a single section, within which are layered further cutaway views penetrating different levels towards the centre of the structure.

VIEWPOINTS AND ANGLES
There is a diagrammatic element to the cutaway: it is a device for providing particular kinds of information,

but that information has to be accurately conveyed and consistent with the other elements explained in the image. As the point of this illustrative convention is to show the internal components in their proper location and in correct relation to the outer layers, this is a considerable factor in selecting the overall viewpoint for the illustration.

The cutaway is logically positioned at the point which reveals most directly the items of interest within the structure, and at the angle which best illustrates the character of the internal parts. But the artist should preserve those elements of the external view which allow the viewer to identify the object. In commercial work, it may also be necessary to incorporate quite clearly an outer section of the object

displaying the manufacturer's name or logo, or the product name. This can eliminate the possibility of employing an unusual angle for dramatic visual effect, as the lettering must not be distorted or made illegible by the angle at which it is displayed. There are many different factors to be taken into account.

DEFINING THE CUT EDGE
When a cutaway is employed to lift out part of a sheet or layer of material, as in the outer casing of a radio or the shell of a boat, the cutout shape has an edge which must be identified as such, and distinguished from edge qualities both on the surface of the object and deeper inside the cutaway section. In some cases the cut edge corresponds to the linear

Engine cross-section
© Saab-Scania
A simple cross-section is a schematic form of cutaway, here treated with airbrushed colour to provide a realistic sense of depth and volume. Central cylindrical elements are shown in full view within the clearly defined section.

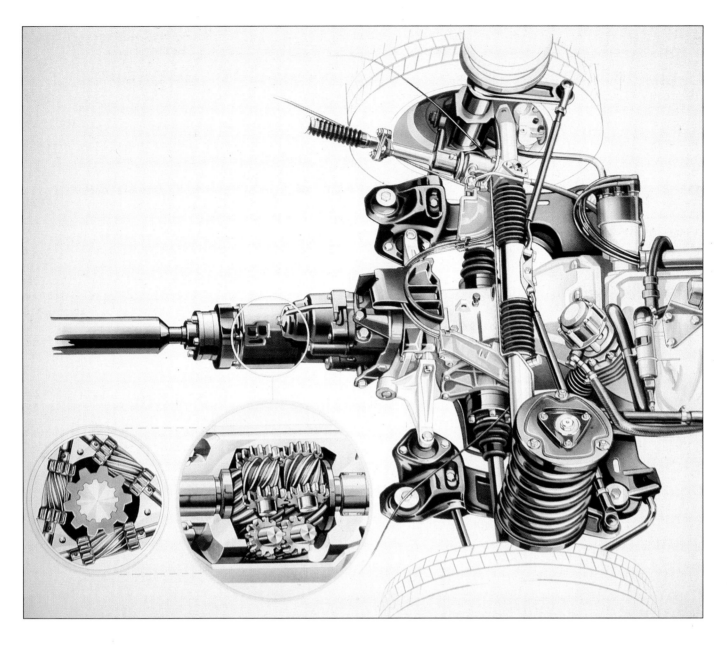

cross-section of the object; alternatively, it may travel through different components, appearing as a continuous line following a fluid or angular path through varying levels and around an assortment of regular or irregular shapes. The edge may be represented smoothly, literally as if cut, or as a wavy or jagged line suggesting a piece broken off.

Where there is an indication of the thickness of the layer of material at the cut edge, this can be represented in different ways. In line work, for example, the depth of the cut edge may be given a linear or finely stippled tone, with surrounding planes left

white; or the edge may be left white against a light texture representing the outer layer of the object. Shadow areas can be defined inside the edge if the cutaway section is hollowed out; this also allows interior components to be clearly represented against the darker tone. Tonal values may be obtained using appropriate techniques of penwork, or they may be laid in with mechanical tints to give a perfectly regular and controlled tonal effect.

In colour rendering, there has been a long-standing convention of colouring the cut edge flatly in flame red, so that it is instantly identifiable. This is

Differential cutaway
© Audi AG
To accommodate the scale of the overall view, the cutaway element is pulled out and presented separately, showing the mechanisms from the angle corresponding to the main image and also in cross-section. Airbrushed colour is used descriptively in the general view and diagrammatically in the cutaway detail.

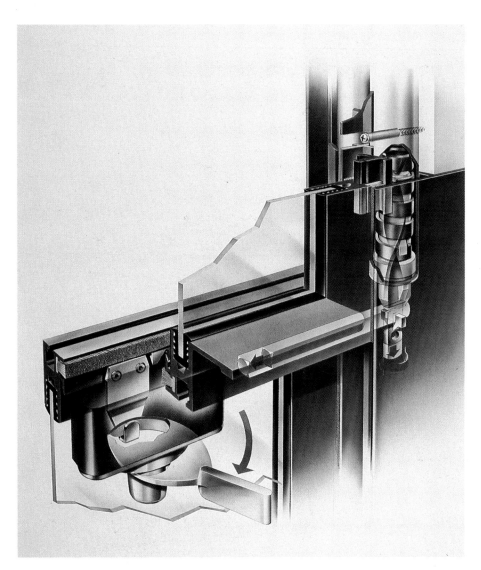

no longer the standard means of representation, but it is still often seen. The use of flat colour to form a broad band around the cutaway is useful in helping the viewer to distinguish this element of the illustration clearly, by contrast with the modelled forms elsewhere in the image. However, the curves and angles of the cut edge are often treated tonally to maintain the overall sense of three-dimensional form. In this case, the edge may be differentiated by selecting a separate colour, or by giving the tonal modelling a directional flow which opposes that of the immediately adjacent areas, drawing attention to the boundary of the cutaway sections.

A section cut into a solid mass of material does not always produce distinctive edge qualities, unless there are several layers of material forming

Above: Window frame
Artist: Hugh Dixon, Spectron Artists
Client: Spectaglaze
A right-angled cut, coloured red, identifies each section of the window frame and locking device. The glass panes have a 'broken-edged' cut indicating continuation of the surfaces. The overall design contrives lively illustration of a relatively mundane technical subject.

Right: 1986 MY 1410I Rolls-Royce engine
Studio: Warwickshire Illustrations Ltd
© Rolls-Royce Motor Cars Ltd
A classic example of the cutaway applied to a classic product, flawlessly rendered with airbrushed colour.

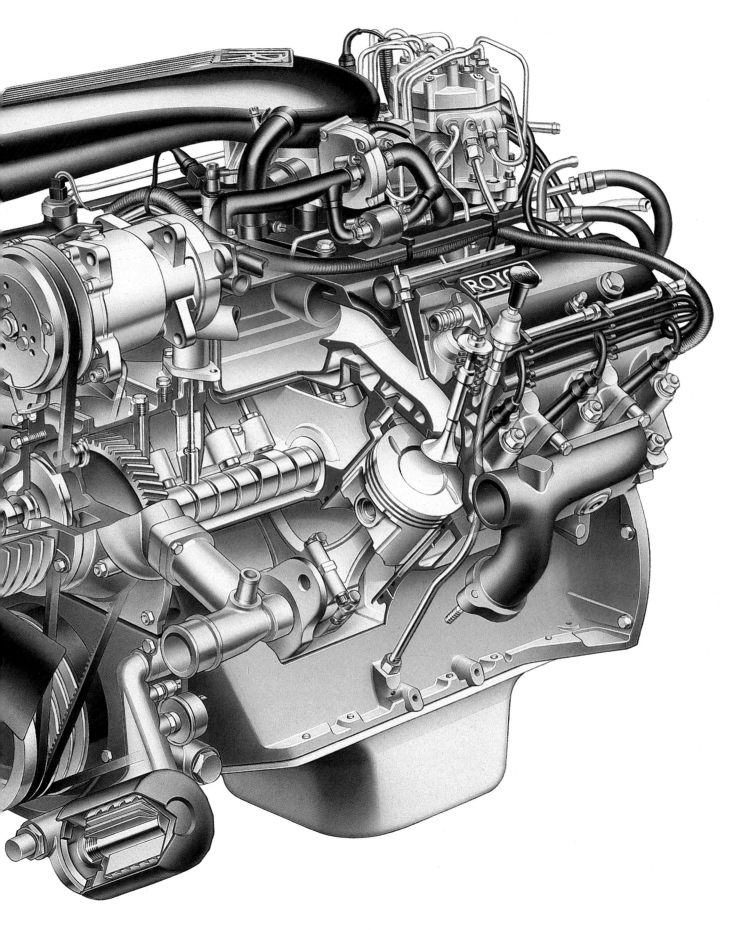

Above: Scorpio key and lock
© Ford Motor Company
**A section cut into a cylindrical
component is a commonly used
visual device because so many
technical objects have a basically
cylindrical form. In this example the
cut is made vertically to the centre of
the lock barrel and angled at more
than 90 degrees, exposing an open
section in which all the details of the
lock mechanism are clearly
described. Since the object is
compact and geometrically logical,
the cutaway rendering can be easily
understood.**

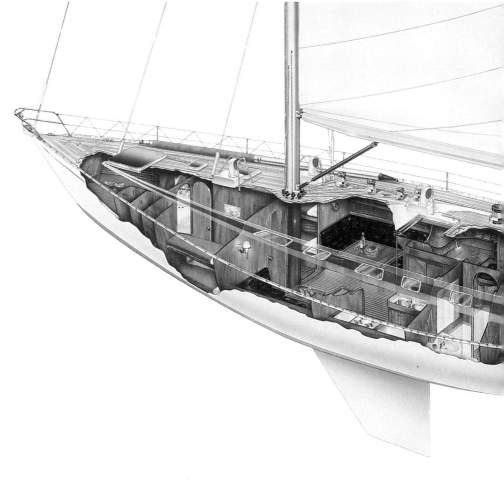

Right: Yacht
Artist: Keith Harmer
Studio: Blue Chip Illustration
Client: Swan Yachts
**An irregular cut edge is combined
here with ghosted elements to reveal
the interior fittings of the yacht
through its exterior shell. The
ghosted surface is necessary to
establish the relationship of the cut
line travelling across the deck and
side of the yacht and also anchors
the perspective of the partitions
below decks.**

Above: 3D camera
Artist: Graham White, Folio
Agency: P.V.A.F.
Client: Nimslo
This is an example of the 'broken-edged' cutaway applied to a small technical object. The line is distinguished by red edged with white where the irregular section of the outer casing and upper surface of the camera are removed. The image is mainly airbrushed in transparent inks to obtain the smooth textures of metal and glass.

the mass. If the material is uniform, the illustrator creates an angled plane which can be treated visually in the same way as the outer surface of the material, manipulating the balance of tone and colour to ensure that the viewer reads the cut section as an interior or receding plane. Where the subject matter is appropriate, there may be integral elements which can be exploited, such as a characteristically different texture where the cut passes through the grain of the material at a different angle. This might be represented as a realistic treatment simulating the actual texture or as a more stylized effect, using

techniques such as hatching or stippling to develop a separate visual identity for the cut surfaces.

A sectioned cutaway view is equally effective in line or full colour. Schematic renderings can also be 'colour coded': without using the full range of representational colour, individual layers or related components are linked and identified by selected colours.

COMBINING CUTAWAYS AND GHOSTING

The alternative method of representing interior functions and components is by ghosting (see following section), in which the outer layers are seen through rather than cut out of the image. It is not uncommon to see both cutaway and ghosted areas in a single image, depending upon which is the most efficient way to penetrate a particular part of an object and explain the internal–external relationships. In recent years, the two illustrative conventions have also been merged in a style which eliminates the cut edge while sometimes leaving interior parts clearly open and untouched by suggestions of surface covering (see pages 6–7).

GHOSTING

Ghosting is a convention used in line work and colour rendering, and particularly well served by airbrushing techniques. Like cutaways, it is employed to reveal internal components and mechanisms, but instead of cutting through the external structure, ghosting renders the outer layers as if they were transparent. A ghosted view of a technical pen, for example, would render the casing as a transparent, basically cylindrical form, within which the reservoir and ink channel leading to the nib are visible. In an open structure, the technique can also be used to make a solid form transparent so that the continuation of the parts passing behind it is clearly indicated. Ghosting is a particularly useful technique in cases where it is essential to retain an impression of the external form; it can also provide more information about the internal detail than does a cutaway view.

INTRODUCING GHOSTED ELEMENTS

In certain forms of technical illustration, ghosting is an essential element providing the viewer with specific information about interior components, machine motions or basic structure. In others, it is designed more to elaborate and authenticate the image, demonstrating the complexity of a machine without instructing the viewer further on the actual functions of the mechanisms. The nature and extent of the ghosted areas often depend upon the intended audience for the work: in an industrial context, the precise construction of internal components or sequential pattern of a moving part may be essential to the person making practical use of the illustration, while in a sales brochure designed for a lay audience, its purpose may be to emphasize the high-quality manufacturing standards applied to the product.

Like cutaways, ghosting can be applied to small sections of an image or may extend almost across the full

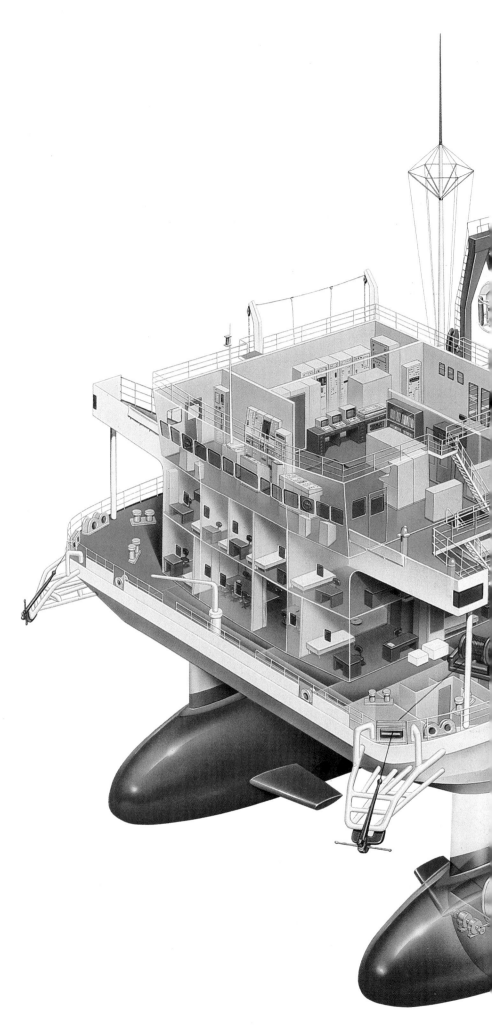

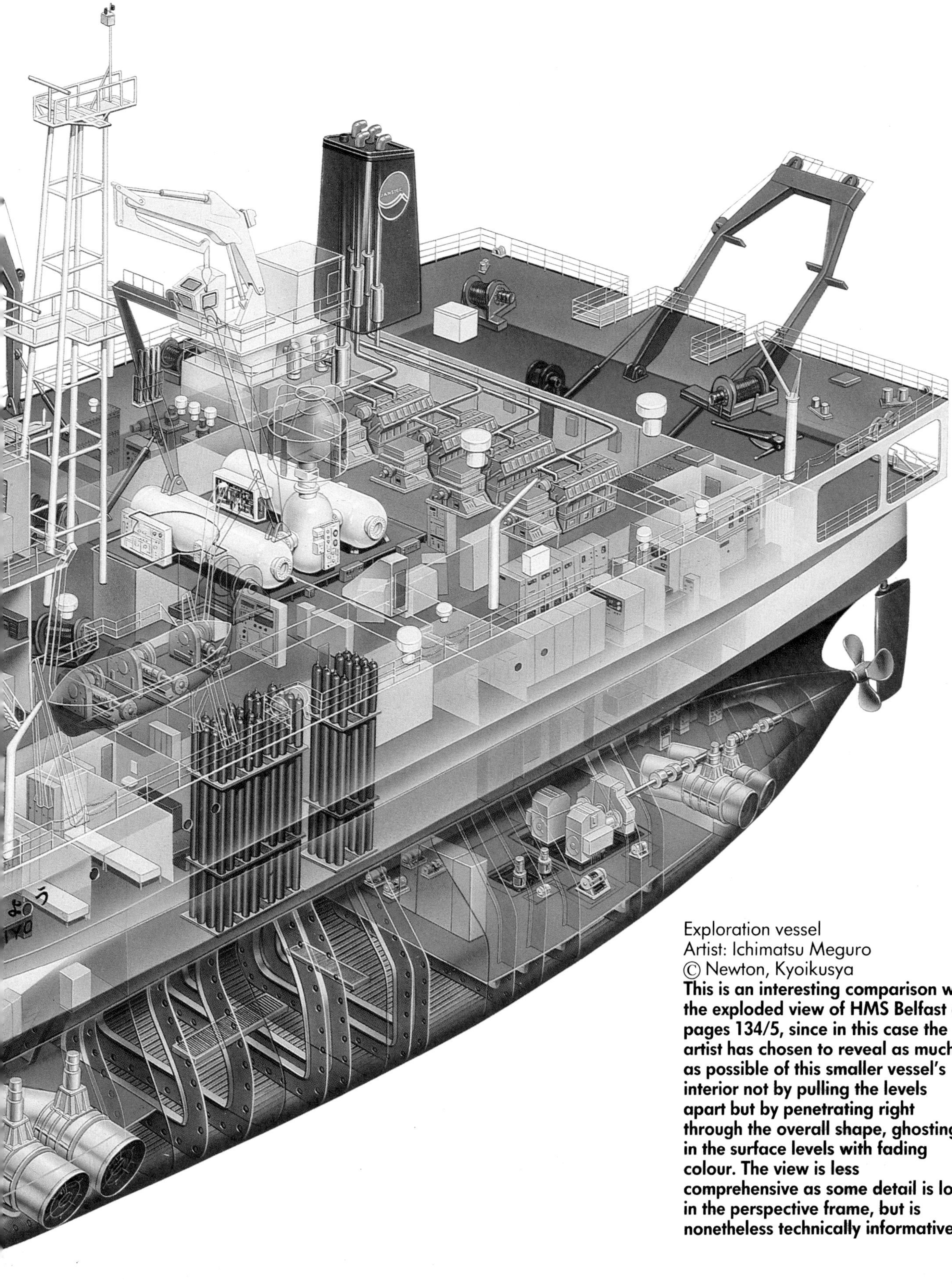

Exploration vessel
Artist: Ichimatsu Meguro
© Newton, Kyoikusya
This is an interesting comparison with the exploded view of HMS Belfast on pages 134/5, since in this case the artist has chosen to reveal as much as possible of this smaller vessel's interior not by pulling the levels apart but by penetrating right through the overall shape, ghosting in the surface levels with fading colour. The view is less comprehensive as some detail is lost in the perspective frame, but is nonetheless technically informative.

Below: Heavy goods vehicle
Artist: Stuart McKay, Aircraft
The viewer's ability to understand the integration of elements in this ghosted truck is assisted by the use of a 'colour coding' system, not unrealistic but sufficiently graphic to identify individual structures and locate them in the construction as a whole. Acrylic paints and inks, applied with an airbrush, create contrasting effects of opaque and transparent colour which form the layered effect of the ghosting.

form, depending upon the range of information to be conveyed. One typical use is to show components fitted beneath a layer of sheet material: for example, the winder mechanism of a camera inside the plastic casing. Some of the most elegantly elaborate ghosted views are applied to products of the automotive industry, such as cars and motorcycles, to reveal what lies beneath the bodywork. The initial viewpoint and the extent of the ghosting are selected to provide the most demonstrative rendering of the object.

Highly sophisticated images can be created by ghosting right through an object from front to back, effectively showing all the interior detail, with just enough of the outer form visible to

identify the object and connect the internal components to the overall design. This is done by a process of layering the image, involving at every stage specific decisions about the amount of detail which shows through one layer into the next, how this can be rendered technically in terms of drawing and painting techniques, and to what extent the construction of successive levels of the component parts can be overlaid and effectively represented without destroying the clarity of the image.

CONVENTIONS OF REPRESENTATION IN GHOSTING

Airbrushing is the painting technique

Rolls-Royce & Bentley wheel designs
Studio: Warwickshire Illustrations Ltd
© Rolls-Royce Motor Cars Ltd
This is an impressive and unusual example of technical illustration in which the ghosting serves not to increase the levels of information but rather to give visual depth and coherence to the overall image. The regularity of the overlapping circles is subtly disrupted by the ghosted shadows and underlying forms which provide fluid links between the different tyre designs. The quality of airbrush spraying is fully exploited to produce the gentle tonal gradations and fine veils of semi-opaque colour.

Ruston portable generator set
Studio: Warwickshire Illustrations Ltd
The outer casing of this portable generator is rendered as completely transparent, but visual clues are retained to establish the external planes through which the interior is seen. Bands of grey and blue passing around the box-like construction maintain the geometric shape. Lettering both identifies the product and locates the surface layer.

most frequently selected for ghosting in colour work, because the quality of the airbrush spray lends itself particularly to conveying transparent effects. Airbrushed colour can be applied so finely that the paint layer itself is actually transparent or translucent. One system of rendering ghosted images therefore consists of working up the internal detail fully before ghosting fine layers of colour over the top to suggest the outer surface planes. Similarly, components which overlap one another as seen from the selected viewpoint are overlaid one on another, using different degrees of transparency and opacity to define different levels of interior depth.

This technique is probably more commonly employed than the alternative approach, which is to render the external form as it is normally seen, and then to ghost the selected ele-ments of the interior over the completed rendering. A delicate approach is needed for this procedure, as it involves drawing up additional detail over previously completed airbrushing, and subsequently applying colour detail which preserves the initial form while giving substance to the ghosted elements.

Where the ghosting identifies a moving part, the same techniques can be applied but the amount of detail in each level may be reversed. The first layer of the rendering may merely establish a plane of flat or graduated tone, the surface against which the component pursues its course of movement. The alternative positions are then ghosted in with transparent and semi-transparent layers. Where the part comes to rest in a fixed position, it is rendered solidly and in full detail. The sequence of work also depends upon the medium being used, keeping in mind that transparent media such as watercolour and ink must build from dark to light areas, while opaque gouache can be overlaid one colour on another to produce either a semi-transparent effect or areas of solid colour obliterating part or all of what has gone before.

The control which the airbrush artist can apply through the systematic process of masking the image to spray different tones and colours, and the subtle qualities of the fine spray of the medium, make airbrushing the obvious choice for ghosted work in full colour. It is more difficult, though not impossible, to achieve the same range of effects applying the colours with a paintbrush, because the simple process of layering the paint surface to correspond to the levels of representation in the image is not so

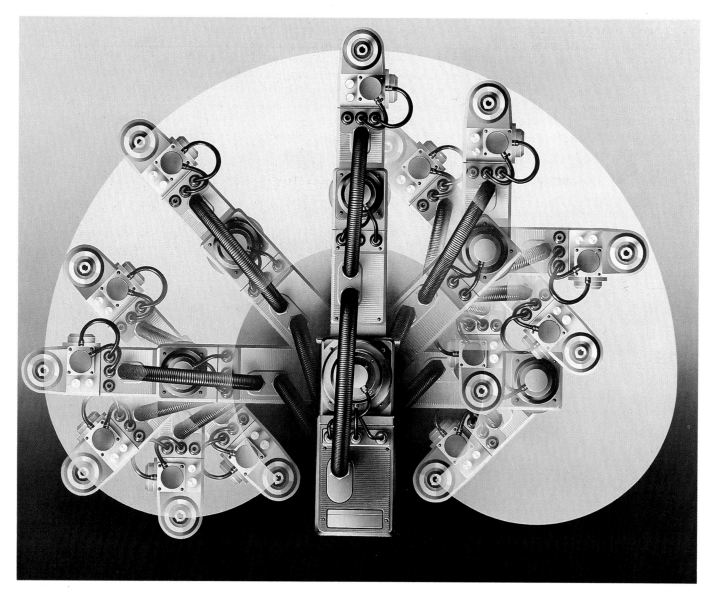

Robot arm
Artist: John Harwood, Folio
Agency: Mayer Norton Group for
Baums Mang and Zimmerman
Client: Bosch
**Ghosted forms demonstrate the
movement of the robot arm from its
central axis and the partial
articulation within the arm. This
makes the technical motion of the
object explicit and visually is a far
more interesting and 'readable'
interpretation than could be
achieved by dealing with the arm as
a continuously solid form.**

easy to carry off. Ghosting with hand-painting techniques requires very careful analysis of colour and tonal values to arrive at the same sort of surface effects by different means. Even in airbrush work, however, there is a lot of fine detail which commonly has to be applied by hand towards the end of the process of rendering. There may be a number of small visual clues, adding to the conviction of the image, which do not lend themselves to airbrushing.

There is a simple but highly effective airbrush technique used to render a translucent surface. A very fine spray of opaque white is applied as a top layer over a detailed rendering: this must be fine enough to be seen through. It suggests the reflective

quality which typically accompanies transparency in a hard material, creating a glassy surface effect.

Other devices for ghosting include bands of translucent colour travelling over the external contours of the form, behind and between which the interior detail is viewed, or simple lines of colour tracing the framework of construction of the object, such as major changes of plane or the junction of interlocking components. Lettering, such as a product or company name or a recognized logo, can be used to establish an otherwise invisible foreground plane, with a faint surround of colour where appropriate to indicate the continuation of the surface layer. The internal elements may be faded off into the solid colour of the overlying plane, which in reality forms a concealing layer, or may be set within a distinctly outlined shape corresponding to a particular component of the overall structure.

The illustrations here and in the preceding pages of this section show various techniques of rendering the ghosted effect.

GHOSTING IN LINE WORK

The principles of applying ghosted detail to line work are the same as for colour rendering, but the restrictions of the technique obviously reduce the range of visual effects the illustrator can employ. The basic ingredients of ghosting here are line weight and broken lines. Broken-line techniques for outlining and shading preserve the consistency of the line weights throughout the image while clearly signalling the different levels of information.

A ghosted section which runs behind an otherwise solid component may be indicated by breaking the line of the ghosting where it crosses the main outlines of the foreground planes. The break should form a demarcation, but the line work needs to remain sufficiently fluid and consistent to allow the viewer to trace the connection of elements on any single level.

Broken-line shading establishes the impression of a transparent foreground plane: the breaking is uneven, created by the technique of 'flicking off' the ink line, causing it to thin and taper away to nothing by speeding up the action of the pen. Stages of movement can be plotted using broken outlines or a line weight clearly different from those of the non-moving parts. Line qualities are selected within the overall context of the drawing, following the same decisions on consistent representation of outline, contour and texture that have been applied to solid elements of the image.

Apple II-e computer
Artist: Dale Gustafson, Mendola Ltd
Two different methods of ghosting are used here to convey specific items of information. The cables connecting the parts of the computer are rendered as substantial elements appearing to pass through the main forms (but note that on the lower right corner of the VDU, the ghosted cable passes behind the product logo leaving this clearly identifiable). Subtle changes of tone along the cable lengths follow the general pattern of colour and tone throughout the rendering. To reveal all the complex detail of the electrical connections and control mechanisms within the keyboard base, the outer casing is lightly ghosted in with fine bands of white which establish the main lines of the exterior shape.

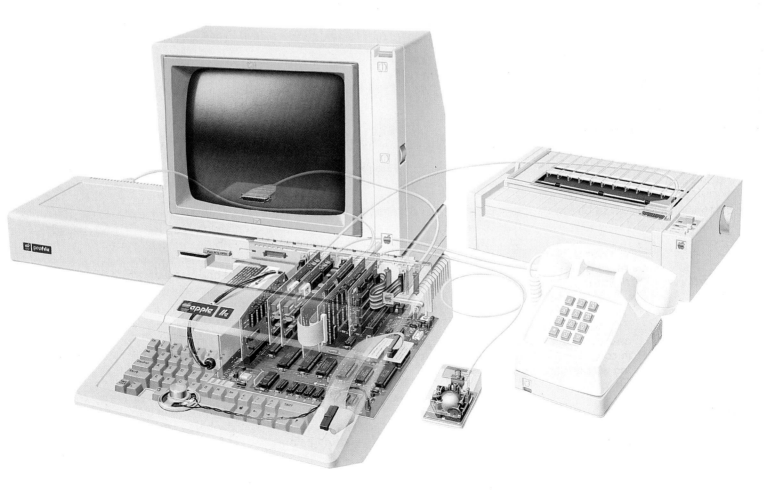

USEFUL ADDRESSES

Simon Adamson
Willow Cottage
25 Station Road
Porchester
Fareham
Hampshire
PO1 68BQ

Aircraft
39–41 New Oxford Street
London
WC1A 1BH

Lloyd Allan
15 Chapelside
Titchfield
Fareham
PO14 4AP

Mark Ansell
18 Langstone Walk
Peel Common
Gosport
Hampshire
PO13 0QW

Pierre d'Avoine Architects
Tapestry Court
Mortlake High Street
London
SW14

Zafer Baran
1 Venn Street
London
SW4 0AZ

Kevin Barnes
No 4 The Spinney
Yately
Camberley
Surrey
GU17 7SJ

Blue Chip Illustration
18 Gloucester Street
Malmesbury
Wiltshire
SN16 0AA

Bournemouth & Poole
College of Art and Design
School of Illustration
Wallisdown Road
Poole
Dorset
BH2 5HH

Colin Brown
454 Bath Road
Saltford
Bristol
BS18 3DJ

Graeme Chambers
108 New River Crescent
London
N13 5RJ

Emma Collins
Flat 4
Woodend
Crawley Ridge
Camberley
Surrey

Timothy Connell
2 Downham Close
Cowplain
Hampshire
PO8 8UD

Philip Crowe
68 Clifton
York
YO3 6AW

CSS Design
Tower House
Southampton Street
London
WC1

Folio
10 Gate Street
Lincoln's Inn Fields
London
WC2

Bob Freeman
77A Conway Road
London
N14

Garden Studio
5 Broad Court
London
WC2 5QH

Michael Gilbert
78 Besley Street
Streatham
London
SW16 6BD

A. Granger
79 Vanner Road
Witney
Oxfordshire
OX8 6LL

Grundy & Northedge
Designers
Thames Wharf Studios
Rainville Road
London
W6 9HA

Dale Gustafson
Mendola Limited
420 Lexington Avenue
New York
NY 10170

Nick Hawken
Higher Trethake
Darite
Liskeard
Cornwall
PL14 5JT

Hop Studios
2 Jamaica Road
London
SE1 2BX

Kevin Jones Associates
12 Marriott Road
Barnet
Hertfordshire
EN5 4NJ

Christopher Kent
47 Wrigglesworth Street
London
SE14 5EG

Steve Latibeaudiere
85 The Avenue
London
N17 6TB

Darren Madgwick
3 Paddington Road
North End
Portsmouth
Hampshire

David Manchip
1A Bollom Flat
Devonshire Avenue
Southsea
Portsmouth
Hampshire
PO4 9EA

Ichimatsu Meguro
2-10-27-30 Komazawa
Setagayaku
Tokyo
Japan

Middlesex Polytechnic
Department of Scientific
and Technical Illustration
Pat Hill
Barnet
Hertfordshire
EN4 8HT

Stuart Michael Molloy
23 Tudor Court
Castleway
Hanworth
Middlesex
TW13 7QQ

Omnific
7–9 Compton Avenue
London
N1 2XD

David Penney
49A Langdon Park Road
London
N6 5PT

Portsmouth College of Art,
Design and Further Education
Technical Illustration Dept
Winston Churchill Avenue
Portsmouth
Hampshire
PO1 2DJ

Jonothan Potter
46 Quebec Street
Brighton
East Sussex
BN2 2UZ

Ravensbourne College of
Design and Communication
Walden Road
Chislehurst
Kent
BR7 5SN

Michael Robinson
71 Park Avenue South
London
N8 8LX

Ron Sandford
Peter Davenport
Associates
116B Newgate Street
London
EC1A 7AE

Roger Savage
26 Montpellier Avenue
Bispham
Blackpool
Lancashire

John Scorey
97A Palmerston Road
Bowes Park
London
N22 4QS

Goro Shimaoka
406–1 Haraichiba
Hannou-Shi
Saitama 357-01
Japan

Spectron Artists
5 Dryden Street
London
WC2E 9NW

Tom Steyer
87 Kirkstall Road
London
SW2 4HE

James Sunderland
G.I.D.
Falmouth School of Art
and Design
Wood Lane
Falmouth
Cornwall
TR11 4RA

Clive David Thomas
20 Redhouse Road
Bodicote
Banbury
Oxfordshire
OX15 4AZ

John Vollands
32 Mountrose Avenue
Intake
Doncaster
South Yorkshire
DN2 6PN

Robert Walster
97A Palmerston Road
Bowes Park
London
N22 4QS

Warwickshire Illustrations
Ltd
45 Blondvil Street
Cheylesmore
Coventry
CV3 5QX

Ross Watton
58 Kinson Grove
Kinson
Northbourne
Bournemouth
Dorset

Brennan Whalley Ltd
131 Kingston Road
London
SW19 1LT

Rosie Whitcher
Science Museum
Exhibition Road
London
SW7

Brian Whitehead
114 Ladbroke Grove
London
W10 5NE

Sean Wilkinson
97A Palmerston Road
Bowes Park
London
N22 4QS

Andrew Wright
15 Edward Barker Road
Heighington
Lincoln
LN7 1TD

Susan Hut Yule
202 West 102 Street 5RE
New York
NY 10025

ACKNOWLEDGEMENTS

Special thanks to:

Ford Motor Co. Ltd
Rolls-Royce Motor Cars Ltd
Staedtler (UK) Ltd
Concord Lighting Ltd
Mobil North Sea Ltd
Mobil Oil Co. Ltd
The Science Museum,
London
Saab-Scania

United Kingdom Atomic
Energy Authority
Zanussi Ltd
Nikon UK Ltd
Wolff Olins
Renault Presse
Lotus
Audi AG

GLOSSARY

Acetate A transparent plastic sheet material, available in various thicknesses, with a number of studio uses, e.g. for masking airbrush work, for colour overlays in illustration.

Acrylics Paints made with a synthetic resin base which can be diluted with water and have the properties of being quick drying and sustaining colour values.

Airbrush An instrument for applying ink or diluted paint in the form of an atomized spray of colour by combining the medium with air under pressure. Characteristically, the colour spread is even and can be subtly graduated.

Artboard A general term for rigid paper materials used as supports for painting and drawing. There is a wide range of weights and textures.

Artwork Original illustration and other types of graphic material presented in a form suitable for reproduction.

Axis An imaginary line representing the linear centre of an object or image. This describes the angle or direction of an object in relation to the viewer in a perspective view, and divides an image symmetrically or with equal balance around the axis.

Background Broadly, the area surrounding the central subject or focal area of an image, which may describe real or apparent space or a specific location.

Bleed 1. In painting, colour which shows through a painted over-layer, or a ragged edge of colour extending over an outline or masked edge. 2. In graphic design, the extension of an image over the outer boundaries of a design grid, e.g. an illustration taken across the margins of a book page.

Brief The instructions given to an illustrator or designer for executing a specific commission. For the illustrator this should specify subject matter, style and the form of presentation required (e.g. line or full-colour), and also the fees and deadlines for the work: it may also include provision of reference material or contacts for technical advice.

Cross-section A diagrammatic or fully realized rendering of an object showing it as if cut right through, usually at right angles to the main axis.

Cutaway An illustration showing an object with a section cut out or with an outer casing partially cut away to reveal interior parts.

Drafting film A semi-transparent sheet material of similar weight to heavy paper, used as a drawing surface particularly for ink line work.

Drybrush A technique of creating a dragged or broken paint texture by charging the brush with colour and blotting it off before applying the residue of paint.

Elevation A diagrammatic view of an object showing a vertical projection of one side only.

Exploded view An illustration showing an object disassembled with the parts aligned in correct relation one to another along the main axes of the construction.

Filling in An effect which may occur in reproduction of line work when printing ink floods the white spaces in delicate line detail and forms a solid black area. This can also occur in printed half-tones and typography.

Foreshortening The perspective effect of recession in which there is apparent distortion of the normal proportions of an object: parts closest to the viewer may appear very large and those furthest away relatively much smaller, and the

distances between them telescoped down by comparison with the actual dimensions.

Format The dimensions and proportions of an illustration, or of the graphic presentation in which it appears, e.g. a book page or wall chart.

Four-colour processing The method of printing in full colour by scanning an image to produce half-tone plates in four colours—yellow, cyan (blue), magenta (red) and black—which when printed together reproduce the original colour effects.

Ghosting The convention in technical illustration of showing interior components or mechanisms of an object as if seen through the outer layers.

Glaze A thin layer of transparent colour. This term usually refers to oil paint diluted with turpentine or oil, but water-diluted acrylics can produce a similar effect.

Gouache Paint consisting of pigment in a gum binder together with a filler substance which gives the colour opacity.

Grain The texture of a paper surface resulting from the fibres in its composition and the method of finishing used in manufacture.

Grid A measuring guide used to structure an illustration or design. A printed or drawn perspective grid may be the basis of an accurate drawing for a technical illustration. In design work, a grid is made to page format and sets the framework for positioning type matter and illustration.

Ground 1. A prepared surface for painting and illustration work. 2. The general area of an illustration on which an object or image is depicted.

Half-tones Patterns of dots or lines simulating solid areas of varying tonal values.

Hatching The method of producing tonal values, mixed colours or textural effects in drawing or painting by filling an area with parallel lines rather than solid or blended colours.

Highlights The lightest tonal or colour values in an image, describing any point or area where the maximum amount of light is reflected from a surface.

Hot-pressed A smooth quality of paper surface created by heat processing in manufacture.

Hue The intrinsic value of a specific colour: for example, red identifies a certain type of colour, but includes a variety of hues ranging from orange-red to purple-red.

Ink A liquid medium for drawing or painting available in black, white and a range of colours. Certain types of drawing ink are formulated to be completely waterproof when dry.

Keyline An outline drawing used as the basis for artwork or illustration.

Layout A rough or finished design showing the positions of component graphic elements such as type and illustration in correct scale and relation to each other.

Line work Illustration drawn with black ink line on white artboard or drafting film. Also, the technique of producing such an image.

Liquid masking Also called masking fluid, a rubber compound solution which is painted on a surface and dries to a thin rubber skin, then acting as a mask until it is lifted and removed.

Local colour The actual colour of an object or surface, disregarding effects of light, shade or reflected colour.

Marker A drawing tool consisting of a metal or plastic casing containing a fibrous reservoir saturated with coloured ink, which feeds a felt or fibre nib.

Mask A material or object used to protect an area of the support while paint or ink is applied, leaving an area uncoloured. This may be paper, board or acetate, an adhesive masking film, liquid masking, or even a three-dimensional object.

Masking film A flexible plastic self-adhesive film used as a mask, particularly in airbrushing but also in association with other painting and drawing techniques.

Mechanical tints Prepared self-adhesive sheets of linear or dot patterns which can be applied to artwork or line illustration to create tonal values.

Medium 1. The material used in painting or drawing, e.g. paint, ink, pastel etc. 2. A substance mixed with paint to give it particular properties, e.g. a glazing medium used with oil or acrylic paints.

Modelling In illustration, the manipulation of colour and tonal values to create an effect of three-dimensional form.

Monochrome The description of a drawing or painting worked wholly in black, white and grey, or in tones of a single colour.

Not Also called cold-pressed, any of various types of moderately smooth paper surface which are not heat-finished.

Objective drawing The practice of drawing from direct observation of an object to produce a detailed and realistic image. Also, a drawing so created.

Oil paints Paints made from pigment mixed with an oil base, which have the properties of being slow drying and retaining colour values.

Opacity The quality of a material which reflects rather than transmits light and cannot be seen through.

Orthographic projection A method of drawing an object by describing it as a series of flat images, in plan view and elevations, which diagrammatically represent the three-dimensional form.

Overlay Part of an illustration or design drawn or painted separately from the main image and presented on a sheet of tracing paper, drafting film or acetate laid over the main image and keyed to it accurately.

Pastel A drawing material consisting of pigment in a gum binder compressed into round or square sectioned sticks.

Perspective The general term for methods of producing accurate two-dimensional representation of three-dimensional form and space, based on the principles of convergence of horizontal lines travelling towards the horizon and diminishing scale in progressively more distant objects.

Picture plane The imaginary surface on which the artist projects a perspective view of a real object. This is usually assumed to be at right angles from the direct line of sight.

Pigments The colouring matter in paints and drawing materials derived from solid natural or synthetic materials finely ground and processed. Liquid dyes are also used as colouring agents.

Plan A diagrammatic view of an object or location as if seen from directly above or below.

PMT Abbreviation for photo-mechanical transfer, the process by which an image can be reproduced, on enlarged or reduced scale if required, by means of a PMT machine designed for graphic purposes. Also, the image so produced.

Reference Material used to provide information for constructing an illustration, which may be in various forms such as engineering drawings, photographs, sketches, objective drawings, written notes etc.

Rendering The process of creating a drawn or painted image. Also, the finished image itself.

Rough 1. A rapid and loosely worked sketch showing the basic form of an illustration or design. 2. Any of various types of paper surface in which the grain and surface texture are relatively coarse.

Scaling up The method of enlarging an image by creating a geometric grid over the original and reworking it on a proportionate grid. The same method can be used to scale down (reduce) the image.

Scratching back A method of altering a painted or drawn image by removing the medium from the support with the blade of a scalpel or similar tool. This is used in airbrushing to remove colour in order to create highlights, and in line work to lift ink where a correction is to be made to the image.

Spattering The technique of applying paint or ink as irregular dots of colour. This can be hand-done by flicking colour from the bristles of a brush, or as an airbrush effect using a special spatter cap or by adjusting the spray quality.

Starburst A highlight effect used in airbrushing which resembles a star radiating light.

Stippling The technique of applying tone or colour as tiny dots rather than solid shading in either painting or drawing.

Study A drawing or painting which forms part of the reference for a finished illustration.

Support The material on which an illustration is painted or drawn, e.g. artboard, paper, drafting film.

Tone, tonal values The range of light and dark elements in a monochrome or colour rendering comparable to a scale of values passing from black through a range of greys to white. Effects of graduations through light and shade are essential to accurate rendering of solid forms.

Tooth The grain or surface texture of a support which enables particles of a drawing or painting medium to attach to it. A smooth artboard has very little tooth; a rough paper, for example, has pronounced tooth.

Trammelling A method of drawing an ellipse by reference to the measurements of its major and minor axes.

Transparency The quality of a material which transmits light and can be seen through.

Underpainting The technique of building up an image by laying in areas of light and shade in monochrome before applying washes or glazes of colour.

Visualizer A machine for projecting a same-size, enlarged or reduced image of a graphic original which can be traced off from a glass screen.

Wash A thin layer of transparent colour created by a paint medium diluted with water. Laying washes of colour is a characteristic technique of watercolour painting but can also be applied to work in gouache or acrylics.

Watercolour Paint consisting of pigment in a gum binder which typically provides transparent colour qualities.

Working drawing A drawing which forms preparation for a finished illustration or design and is produced as part of the working process rather than as an image in its own right.

INDEX

Italic type indicates illustrations